Teaching College Geography

A Practical Guide for Graduate Students and Early Career Faculty

EDITORS

Michael Solem, *Association of American Geographers*
Kenneth Foote, *University of Colorado at Boulder*

Upper Saddle River, NJ 07458

Library of Congress Cataloging-in-Publication Data
Teaching college geography : a practical guide for graduate students and early career faculty/editors, Michael Solem, Kenneth Foote. — 1st ed.
p. cm.
ISBN-13: 978-0-13-605447-4
ISBN-10: 0-13-605447-1
1. Geography,—Study and teaching (Higher) I. Solem, Michael. II. Foote, Kenneth.
G74.T44 2009
910.71'1—dc22
2008007419

Editor-in-Chief, Science: Nicole Folchetti
Publisher, Geosciences and Environment: Dan Kaveney
Project Manager: Tim Flem
Marketing Manager: Amy Porubsky
Production Manager: Kathy Sleys
Creative Director: Jayne Conte
Cover Design: Maureen Eide
Cover Illustration/Photo: Beth Schlemper and Doug Gamble
Full-Service Project Management/Composition: Suganya Karuppasamy/GGS Book Services
Printer/Binder: Courier Companies, Inc.

This material is based upon work supported by the National Science Foundation under Grant No. REC-0439914 and DUE 0089434. Any opinions, findings and conclusions, or recommendations expressed in this material are those of the author(s) and do not necessarily reflect the views of the National Science Foundation (NSF).

Credits and acknowledgments borrowed from other sources and reproduced, with permission, in this textbook appear on appropriate page within text.

Pearson Education Ltd., London
Pearson Education Singapore, Pte. Ltd
Pearson Education Canada, Inc.
Pearson Education–Japan
Pearson Education Australia PTY, Limited
Pearson Education North Asia, Ltd., Hong Kong
Pearson Educación de Mexico, S.A. de C.V.
Pearson Education Malaysia, Pte. Ltd.
Pearson Education Upper Saddle River, New Jersey

10 9 8 7 6 5 4 3 2 1
ISBN-13: 978-0-13- 605447-4
ISBN-10: 0-13-605447-1

CONTENTS

ASPIRING ACADEMICS AND *TEACHING COLLEGE GEOGRAPHY:* A COMMUNITY-BASED WEB SITE FOR NEW SCHOLARS

The companion website for this book (www.pearsonhighered.com/aag) contains a wealth of additional resources for professional development and community building.

Each chapter is supported by activities for workshops, courses, seminars, brown bags, and informal gatherings among graduate students and faculty. The activities provide procedures, recommended readings, worksheets, and other supplementary materials that can help you get started with applying new skills and ideas in your own professional practice.

The website also features tools for sharing ideas and effective practices with others. Use the discussion boards to exchange perspectives and advice about professional development issues with the larger community. Alternatively, you can set up private forums for groups organized by teaching assignment, academic department, research network, or other interest area.

You can also participate in the development of the website by contributing your own classroom activities, links, and professional development resources.

We hope you find the website to be a valuable complement to the text.

PREFACE

Steps to Becoming a Critically Reflective Teacher

Michael Solem and Kenneth Foote

This book and its companion, *Aspiring Academics* (Solem, Foote, and Monk 2009), are part of a multi-year effort to improve the preparation of faculty, instructors, and teaching assistants (TAs) entering the field of geography. If you are just getting started in your career, *Teaching College Geography* provides advice you can apply immediately to your courses and introduces a range of topics to explore further as you develop and refine your skills. We also think this text can relieve some of the anxiety that you may experience as a new teacher and help make teaching an enjoyable and rewarding experience.

Although many graduate programs in geography provide some sort of teaching apprenticeship, a large proportion of faculty enter the academy with relatively little preparation in teaching and learning issues (Solem and Foote 2004). As a result they rely on intuition and their own experiences when facing students for the first time. While this works well for some, other new instructors report an overwhelming feeling of helplessness as they scramble to prepare lectures, grade assignments, and learn the ins and outs of new classroom technologies, all the while fielding e-mail from anxious students worried about upcoming exams or assignments. Other obligations can be just as pressing, so it is no small wonder that teaching is cited by so many early career faculty as one of the most difficult and worrisome aspects of their working lives (Boice 2000; Wulff and Austin 2004).

The good news is that when teaching preparation is addressed early through professional development, much of that stress is quickly replaced by creativity and enjoyment (Nyquist et al. 1999). While graduate school is often the place to deepen and extend our subject matter expertise and research skills, one can also develop a repertoire of teaching methods and explore approaches to student learning. Then, as new faculty gain experience on the job, they can reflect on their experiences; experiment with other more demanding strategies involving field study, inquiry learning, collaborative learning, geographic information systems (GIS), and digital mapping; and gradually develop a thorough expertise in the areas of learning and teaching.

We also believe that teaching should be a social, collegial pursuit, one valued and supported by colleagues and institutions. *Teaching College*

Geography places significant emphasis on creating a departmental climate that values teaching quality and improves the professional development of graduate students and early career faculty. Becoming an effective college and university instructor should be seen as more than an individual responsibility. We should open up our classrooms to colleagues and peers, share ideas, and exchange strategies. Too often course preparation is seen as solitary work to be shared only with students when in a classroom. This is unfortunate because departments themselves play an important role in advancing or undermining the achievement and performance of young faculty in the college classroom. Research has shown repeatedly that when graduate students and early career faculty receive mentoring and professional development in teaching from a community of peers, advisors, and senior colleagues, they are better positioned for success and establishing themselves as confident, effective teachers when they enter an academic position. This, in turn, makes it easier for new teachers to manage all other aspects of their professional lives in the academy (Fink 1984; Boice 1992, 2000; Sorcinelli and Austin 1992; Marincovich, Prostko, and Stout 1998).

Our goal with this book, therefore, is to make more widely available those practices that are known to be effective in helping new instructors thrive in the modern college classroom. We think the following chapters will eliminate much of the guesswork involved with getting started in teaching and help readers build confidence in their abilities. Just as importantly, we hope this book will encourage reflective dialogue within departments—conversations about teaching that recognize and celebrate the talents and perspectives of both new and experienced faculty. Because no department can survive without healthy undergraduate enrollments, we concur with many colleagues (*e.g.*, Halseth and Fondahl 1998; Estaville, Brown, and Caldwell 2006) that it is the collective responsibility of faculty and departmental leadership to ensure that the courses and educational experiences offered in the undergraduate curriculum are of the highest academic quality. The secret to this achievement lies in the professional development of graduate TAs, course instructors, and early career faculty teaching those courses.

Our vision of teaching also rests on the view that the many responsibilities of academic life must be considered together, in balance, and that it is important for us to understand how teaching, research, service, outreach, and our personal lives intersect and interconnect. As in *Aspiring Academics*, we attempt to take a broader view of professional life, one based on Boice's (1992, 2000) vision of moderation, balance, and the seamless, positive integration of the diverse elements of life and work. We stress issues of teaching here in *Teaching College Geography*, but these always should be set in the context of the broader issues of academic life—work–life balance, career planning, and collegiality—which we believe are also keys to long-term success and career satisfaction. Achieving this balance involves adopting the conception of teaching as scholarly work, a view that lies at the heart of the international movement inspired by Boyer (1990) to promote the scholarship of teaching and learning (SoTL) (Healey 2003). The goal of SoTL is:

> to promote significant, long-lasting learning for all students as well as to reward and recognize effective teaching in the same way as other types of scholarly work. SoTL implies that we apply the same standards of rigor we use in our scholarship and scientific research to our approaches to teaching and learning. In part this means opening up our teaching for discussion, feedback, review and collaboration among colleagues, peers, and students—just as we would do with our research —rather than seeing teaching as something private and confined to the classroom. This also means that we can strive to create more effective linkages between our teaching and research and, indeed, actually researching our teaching, teaching materials, and teaching strategies to see whether they are effective—and disseminating the results in print and in public presentations. (Solem, Foote, and Monk 2009)

SoTL stresses the importance—and provides the means—for you to become a critically reflective teacher who always carefully considers the effectiveness of your practices, keeps abreast of research on teaching and learning, and contributes to this process of improvement in your department, university, and discipline.

USING THIS BOOK

In the interest of making teaching more of a community-based, social activity, we have designed this book for groups of graduate students and faculty with active teaching assignments. Individual readers will gain much from the chapters, but we greatly encourage reflection and discussion among peers and colleagues. Section I, "Getting Started," is intended as a standalone introduction to teaching, with readings and activities to support discussion over an academic quarter or semester. This section is based on the materials originally developed for a one-credit course for new TAs by Teresa Dawson, then at Pennsylvania State University, and now director of the Teaching and Learning Centre at the University of Victoria. We have worked with Teresa to update and modify her chapters for a broader audience of new course instructors and professors.

Section I focuses on the nuts-and-bolts issues faced by any first-time teacher: how to design a course syllabus, maintain order and respect in the classroom, prepare students for small-group work, and establish clear and fair grading policies. These topics are sequenced so readers can build upon their skills from one week to the next. Ideally, you should complete the activities in Section I in meetings with peers in your department, and perhaps also with experienced graduate students, advisors, and faculty mentors. If you are working through the chapters on your own, we encourage you to use the online discussion forums featured on this book's web site to share ideas and discuss issues with others.

Once you are familiar with the material in Section I, you can begin to extend your professional development by exploring Section II, "Promoting a Scholarship of Teaching and Learning in Geography." Each chapter in Section II is designed to develop your knowledge and skills in areas of special interest to geography educators: engaging learners in large-enrollment geography classes; supporting critical thinking and analysis with GIS and mapping technologies; designing field studies in human and physical geography courses; and creating learning experiences that foster international perspectives on geographic issues. Although the chapters focus on different topics, all of the authors ground their recommendations in the scholarly research literature on teaching and learning. In each chapter, you will find sidebars highlighting examples of relevant research on student learning, approaches to project evaluation, and ideas for designing a study for assessing the impact of a new student activity. Each chapter, furthermore, features an activity on this book's web site that can help you immediately apply new ideas in your own courses.

We hope that, by completing the activities in each chapter, you will feel more confident about setting course goals, designing a syllabus, creating a classroom presentation, writing a new active-learning or inquiry-based activity, and creating new assessment instruments. After a period of practice and reflection, perhaps you will feel inspired to design an educational research project, publish in an academic journal, or present the results of your teaching innovations at a professional conference—all of which are essential to promoting SoTL in geography. By taking such steps at this early stage of your career, you will play an important role in enhancing the quality of geography in higher education.

A DISCIPLINE-BASED APPROACH TO PROFESSIONAL DEVELOPMENT

This book and *Aspiring Academics* (Solem, Foote, and Monk 2009) are two major outcomes of "Enhancing Departments and Graduate Education (EDGE) in Geography," a project led since 2005 by the Association of American Geographers (AAG) with funding from the National Science Foundation (NSF grant REC-0439914). But our concern for issues of professional development in geography goes back further to other NSF projects beginning with the Virtual Geography Department Project between 1996 and 1998 and the summer institutes organized annually since 2002 for the Geography Faculty Development Alliance at the University of Colorado at

Boulder. Workshops were key components of these projects and were designed to provide faculty with the training needed to develop effective instructional materials and methods for teaching geography (Solem 2000; Solem and Foote 2004).

As workshop facilitators in these projects, we quickly realized that, despite their enthusiasm for engaging with teaching and learning issues, a great number of participants had never been exposed to the theoretical and practical expertise in pedagogy that would allow them to get the most out of their efforts. Certainly these participants were excellent, even award-winning professors with a good deal of practical experience in the classroom, but they had not had exposure to topics relating to instructional materials development, learning theories, principles of assessment, or methods of course evaluation. Many also had some exposure to issues of learning and teaching while in graduate school. But the topics they covered then, and the skills they mastered in the past, simply did not match what they wanted to do in the future. Reflecting on these observations, it seemed clear that for geographers to face future challenges and opportunities, the discipline needs to provide better professional support as early as possible for graduate students and new faculty.

We hope this book will help support this improved training for early career faculty and graduate students. And though this book focuses on geography, readers will also want to consult other excellent interdisciplinary sources. We especially recommend McKeachie and Svinicki (2005), Fink (2003), Davis (1993), Biggs (1999), and Angelo and Cross (1993). Also of great value are Walvoord and Anderson (1998), Wiggins (1998), and Wiggins and McTighe (2005). For a better understanding of learning theory, a number of books have been published for college teachers in the last decade or so that address this issue (Bransford 1999; Zull 2002; Svinicki 2004).

However, there are topics and skills that, though not unique to geography, are best addressed in discipline-specific books like this one. This is particularly true of the emphasis geographers place on field study and fieldwork; the value they place on developing cartographic skills; the importance they see in implementing advanced technologies such as GIS, remote sensing, and Global Positioning Systems in the college curriculum; the fundamentally strong relationship between geography and international education; and the innovative ways they can use the new telecommunications technologies such as the World Wide Web in the classroom and laboratory. You will find all of these topics addressed in this book.

Faculty can gain much discipline-specific training like this, but equally important are campus-based and interdisciplinary programs aimed at improving the instructional skills of graduate students and faculty across many disciplines. Were it not for the literature contributed by faculty development professionals and educational researchers, this disciplinary resource would not be possible. Recognizing the inherent value of this expertise, *Teaching College Geography* was developed with input from scholars affiliated with the Professional and Organizational Development (POD) Network and the Carnegie Academy for the Scholarship of Teaching and Learning (CASTL) Leadership Program.

POD is the primary professional organization in North America for the field of faculty and instructional development. Its mission is to promote a deeper understanding of high-quality teaching and learning and to promote a utilization of that understanding throughout higher education. The CASTL Leadership Program, in which the AAG participates, allows geographers to share the results of their teaching and course practices with colleagues around the world. As one example, the community-based forums on this book's web site provide a place for you and other readers to share ideas, knowledge, and experience. This type of activity is at the heart of SoTL. In the spirit of a participatory, broad-based approach to professional development, we circulated drafts of *Teaching College Geography* to nine academic geography departments for review by teams of graduate students and faculty. The result, we think, is a stronger, more readable book offering a wealth of carefully tested materials.

ACKNOWLEDGMENTS

This book has been a pleasure to work on, due in large part to the generosity of the many authors and reviewers. There seems to be a widespread recognition that more and better professional development is vital to geography—that the time has come to reshape and reform the discipline in ways that will maintain its vitality long into the future.

We wish to thank in particular the advisors, research assistants, and volunteers who served for multiple years on the EDGE and GFDA projects: Rachelle Brooks, Ivan Cheung, Tim Conroy, Teresa Dawson, Dee Fink, Robin Friedman, Gayathri Gopiram, J. W. Harrington, Jr., Matt Koeppe, Vicky Lawson, Jenny Lee, Jongwon Lee, Duane Nellis, Mark Purdy, Doug Richardson, Sue Roberts, Fred Shelley, Beth Schlemper, Patricia Solis, Adam Thocher, and Antoinette WinklerPrins. Myles Boylan at the National Science Foundation provided valuable advice on many occasions on these and other projects.

We also acknowledge the important service of the following professors and graduate students who reviewed and tested early drafts of the chapters and activities: Anneliese Vance, Alan MacPherson, Li Li, Suzanne M. Roussie, and Trina Hamilton (SUNY-Buffalo); Kathy Schroeder, Gabrielle L. Katz, and Christopher A. Badurek (Appalachian State University); Joy Fritschle, Alex Vias, Melinda Daniels, Melanie Rapino, and Ben Franek (University of Connecticut); Joel W. Helmer, Brad Bays, Stephen O'Connell, and Brett Chloupek (Oklahoma State University); Ed Jackiewicz, John P. Salapa, and Ronald A. Davidson (California State University–Northridge); Kate A. Berry, Jill Heaton, Matthew Fockler, Kathleen Poncy, and Timothy Weigel (University of Nevada–Reno); Giorgos Mountrakis, Linda J. Quackenbush, Jida Wang, and Yvonne E. Paul (SUNY–College of Environmental Science and

Forestry); Thomas M. Whitmore, Margaret Carrel, Joseph Santos Escobedo Palis, Elisabeth Root, and Monica Lipscomb Smith (University of North Carolina–Chapel Hill).

References

Angelo, T. A., and K. P. Cross. 1993. *Classroom assessment techniques: A handbook for college teachers*. San Francisco: Jossey-Bass.

Biggs, J. 1999. *Teaching for quality learning at university: What the student does*. Buckingham, U.K.: Open University Press.

Boice, R. 1992. *The new faculty member: Supporting and fostering professional development*. San Francisco: Jossey-Bass.

———. 2000. *Advice for new faculty members*. Needham Heights, MA: Allyn and Bacon.

Boyer, E. L. 1990. *Scholarship reconsidered: Priorities of the professoriate*. Princeton, NJ: Carnegie Foundation for the Advancement of Teaching.

Bransford, J. D., A. L. Brown, and R. R. Cocking, eds. 1999. *How people learn: Brain, mind, experience, and school*. Washington, DC: National Academy Press.

Davis, B. G. 1993. *Tools for teaching*. San Francisco: Jossey-Bass.

Estaville, L., B. Brown, and S. Caldwell. 2006. Geography undergraduate program essentials: Recruitment. *Journal of Geography* 105(January/February): 3–12.

Fink, L. D. 1984. *The first year of college teaching*. San Francisco: Jossey-Bass.

———. 2003. *Creating significant learning experiences: An integrated approach to designing college courses*. San Francisco: Jossey-Bass.

Halseth, G., and G. Fondahl. 1998. Re-situating regional geography in an undergraduate curriculum: An example from a new university. *Journal of Geography in Higher Education* 22(3):335–45.

Healey, M. 2003. Promoting lifelong professional development in geography education: International perspectives on developing the scholarship of teaching and learning in higher education in the 21st century. *The Professional Geographer* 55(1):1–17.

Marincovich, M., J. Prostko, and F. Stout, eds. *The professional development of graduate teaching assistants*. Bolton, MA: Anker Publishing.

McKeachie, W. J., and M. Svinicki. 2005. *McKeachie's teaching tips: Strategies, research, and theory for college and university teachers*, 12th ed. Boston: Houghton Mifflin Co.

Nyquist, J., L. Manning, D. Wulff, A. Austin, J. Sprague, P. Fraser, C. Calcagno, and B. Woodford. 1999. On the road to becoming a professor: The graduate student experience. *Change* May/June:18–27.

Solem, M. 2000. The virtual geography department: Assessing an agent of change in geography education. *Journal of Geography in Higher Education* 24(3):353–64.

Solem, M., and K. Foote. 2004. Concerns, attitudes, and abilities of early-career geography faculty. *Annals of the Association of American Geographers* 94(4):889–912.

Solem, M., K. Foote, and J. Monk, eds. 2009. *Aspiring academics: A resource book for graduate students and early career faculty*. Upper Saddle River, NJ: Prentice Hall.

Sorcinelli, M., and A. Austin, eds. 1992. *Developing new and junior faculty*. San Francisco: Jossey-Bass.

Svinicki, M. D. 2004. *Learning and motivation in the postsecondary classroom*. Bolton, MA: Anker Publishing.

Walvoord, B. E., and V. J. Anderson. 1998. *Effective grading: A tool for learning and assessment*. San Francisco, CA: Jossey-Bass.
Wiggins, G. 1998. *Educative assessment: Designing assessments to inform and improve student performance*. San Francisco: Jossey-Bass.
Wiggins, G., and J. McTighe. 2005. *Understand by design*, 2nd. ed. Upper Saddle River, NJ: Pearson Education, Inc.
Wulff, D., and A. Austin, eds. 2004. *Paths to the professoriate: Strategies for enriching the preparation of future faculty*. San Francisco: Jossey-Bass.
Zull, J. E. 2002. *The art of changing the brain: Enriching the practice of teaching by exploring the biology of learning*. Sterling, VA: Stylus.

ABOUT THE EDITORS

Michael Solem is Educational Affairs Director for the Association of American Geographers, where he directs the Enhancing Departments and Graduate Education in Geography (EDGE) project and the Center for Global Geography Education initiative funded by NSF. Dr. Solem is the external evaluator for the Geography Faculty Development Alliance and the Graduate Ethics Education for Future Geospatial Technology Professionals projects. He currently serves as the North American coordinator of the International Network for Learning and Teaching Geography in Higher Education (INLT), is associate director of the Grosvenor Center for Geographic Education at Texas State University-San Marcos, and leads the AAG's efforts with the Carnegie Academy for the Scholarship of Teaching and Learning program. He received the *Journal of Geography in Higher Education's* biennial award for promoting excellence in teaching and learning for his research with Ken Foote on faculty development in postsecondary geography.

Kenneth Foote is professor of geography and former department chair at the University of Colorado at Boulder. His research spans cultural and historical geography, GIScience and cartography, and geography education, especially the application of instructional technologies and issues of geography in higher education. Ken has led the NSF-funded Geography Faculty Development Alliance since 2002 and is co-principal investigator of the Enhancing Departments and Graduate Education in Geography project with Michael Solem and Janice Monk. He has served as president of the National Council for Geographic Education (2006) and national councilor of the Association of American Geographers (2002–2005). Ken received the Association of American Geographers' 1998 J. B. Jackson Prize for his book *Shadowed Ground: America's Landscapes of Violence and Tragedy* and the association's 2005 Gilbert Grosvenor Honors in Geographic Education.

ABOUT THE AUTHORS

Teresa Dawson directs the Teaching and Learning Centre at the University of Victoria. In 2004 she helped draft the University of Toronto at Scarborough's new official teaching guidelines, which now document concrete ways to value and assess various aspects of teaching—including the scholarship of teaching and learning (SoTL)—for purposes of tenure and promotion. She has been instrumental in founding an Institute for the Scholarship of Teaching and Learning in the Disciplines at UTSC. Nationally, Teresa is a member of the steering committee for the Society for Teaching and Learning in Higher Education (STLHE). Internationally, she participates in the scholarship of teaching and learning both within her own discipline (for example, as a consultant to the AAG) and in collaboration with other SoTL colleagues in the U.S., U.K., and Asia.

Douglas W. Gamble is an associate professor of geography and the director of the Laboratory for Applied Climate Research in the Department of Geography and Geology, University of North Carolina at Wilmington. He has taught a wide variety of physical and environmental geography courses during his career and at least one course each semester for the past thirteen years has had an enrollment of over one hundred students. Dr. Gamble also frequently takes students to San Salvador, Bahamas to participate in field courses and research. During the summers, he teaches a field course for K-12 teachers on the Outer Banks. He has received the University of Georgia Outstanding Teaching Assistant Award (1993) and the NCGE Distinguished Teaching Achievement Award (2005). In 2004 and 2005 he was a co-leader of the Geography Faculty Development Alliance workshops at University of Colorado, Boulder. Dr. Gamble's current research activities include the hydroclimatology of the Caribbean, weather patterns and rainwater chemistry in the Wilmington area, and flood monitoring along the southeastern North Carolina coast. He has received funding for his research from the Association of American Geographers, the National Science Foundation, and the NOAA Coastal Services Center, resulting in publications in *Climate Research, International Journal of Climatology, Journal of Coastal Research, Physical Geography,* and *Theoretical and Applied Climatology.* Nearly half of his publications are coauthored with students.

Cary Komoto is a professor of geography at the University of Wisconsin–Barron County in Rice Lake, Wisconsin, where he began as an assistant professor in 1991. He is currently the chair of the University of Wisconsin Colleges Department of Geography and Geology. Years ago Cary began engaging in the Scholarship of Teaching and Learning (SoTL) and has collaborated on a number of SoTL projects. Some of his most recent projects include one with several department-member colleagues to correct common misconceptions in physical geography and geology. He is collaborating with three colleagues in other disciplines to investigate what happens when students engage in metacognitive reflection on their own and their classmates' learning in diversity courses. Along with four other colleagues he is developing a lesson to help correct student confusion over global warming and ozone layer depletion.

Richard B. Schultz is a nationally certified geologist, GIS Certificate Program Administrator, and researcher in the areas of online (distance) learning, geography and GIS education, and the geochemistry of economic deposits. He specializes in the infusion of spatial concepts and technology into the higher education curriculum, and applying geographic concepts into various cultural settings. Dr. Schultz has written approximately forty manuscripts, book chapters, and articles for international publications. His list of awards includes educational excellence, outstanding teaching awards, and business ethics awards. Rich is a faculty member in the Department of Geography and Geosciences at Elmhurst College in suburban Chicago, Illinois, where he teaches courses in meteorology, physical geography, GIS, intercultural studies, and environmental issues. He is currently the chair for the AAG Committee on College Geography and Careers.

Cathryn E. Springer is an assistant professor at Southern Illinois University Edwardsville in the Department of Geography and the Office of Science and Mathematics Education. She teaches and develops courses in geography and geoscience education and has research interests in virtual field studies, place-based educational theory, and pedagogical fieldwork.

Diana Stuart Sinton is the director of Spatial Curriculum and Research at the University of Redlands in California, where she advises, teaches, and writes about the roles of GIS, mapping, and spatial thinking in higher education. She is an author and coeditor of *Understanding Place: GIS and Mapping Across the Curriculum* (ESRI Press, 2007). Previously she directed the GIS and mapping initiative at the National Institute for Technology and Liberal Education (NITLE), where she promoted the use of mapping and GIS to enhance spatial understanding through a series of workshops, curricular materials, and web-based resources. She is a past recipient of an International Research Fellowship from the National Science Foundation. She taught landscape ecology, geography, and GIS at the University of Rhode Island and Alfred University, and has conducted research in the Pacific Northwest and Argentina. She has degrees in geography from Oregon State University (Ph.D., 1996; MS, 1992) and in religion from Middlebury College (BA, 1988).

Jennifer Speights-Binet is an assistant professor of geography at the University of Houston, Clear Lake. For the past four years, she has been co-instructor of a popular Geography of Texas field trip which, along with her commitment to incorporating field study into all classes, has prompted a research agenda addressing the challenges and benefits of designing geographically meaningful field experiences. Additionally, she is project associate for a National Geographic Education Foundation Grant, entitled "Growing Teachers in Texas Soil." By training, Jennifer is a cultural geographer particularly interested in urban revitalization, New Urbanism, and landscapes of memory and nostalgia.

SECTION I

Getting Started

Teresa Dawson, Kenneth Foote, and Michael Solem

Every geography department has its excellent teachers. One may be known for his or her dynamism in the classroom and wonderful field trips. Another may have been instrumental in fostering diversity in the classroom. Yet another may be loved by students for his or her care and concern for their progress and attention to detail in their learning. By the end of your time in your department you will have your favorites. Each is likely to be different in teaching style, personality, and method, yet they will all be effective teachers.

The point is that there are no rules for being a great teacher. Each of us finds our own style, and the most successful are probably those who are most comfortable with the style they find. You remember the first time you wore a pair of jeans that really fitted, not hand-me-downs, but your own—they just felt right. The more you wore them, the better they felt and adjusted to you until pretty soon you forgot you had them on. Teaching is just like that. Some of us find the right pair the first time; others of us seem to try on a hundred pairs before one fits. We do, however, think that everyone can benefit from some advice to make the process less stressful and more enjoyable.

This first section of *Teaching College Geography* is designed as a stand-alone guide for new teachers, whether graduate teaching assistants, lecturers, adjunct faculty, or early career tenure-track professors. The eight chapters are designed to be covered in eight meetings that you can either conduct individually with your advisor or mentor, or informally as a series of meetings with a cohort of teachers in your department. Alternatively, your department might use these chapters as a resource supporting a more formal professional development course. Whatever the format or approach, by using this guide and completing the activities during an academic quarter or semester when you are teaching a geography course, you can share your classroom experiences with peers and mentors while you simultaneously develop your teaching skills.

The first two meetings should occur in the week or two before classes begin, with the rest of the meetings spaced at appropriate intervals thereafter. The meetings and activities are built on the experiences of past teachers and therefore are structured to meet your needs as much as possible. The chapters will integrate and provide support for the variety of classroom experiences that you are likely to encounter during your first semester or term. For example, there is a progression from dealing with the nuts and bolts of that first classroom situation to reflecting on the more theoretical issues such as student learning styles. Pertinent topics have been scheduled when experience suggests they are likely to come up. There are ten activities included in this section, half designed for the group meetings with a facilitator and half for individual work and reflection.

Most importantly, we hope that these combined resources will convince you that you do not have to manage alone. The key to thriving as a teacher is to learn to communicate. With this in mind, the goal of these chapters, beginning in the first meeting, is to foster an environment for the sharing of ideas and concerns about teaching (and academic life in general) that should stand you in good stead for the rest of your academic teaching career. From there, you can explore the four chapters that form the remainder of the book for more in-depth treatment of teaching issues that are commonly encountered in geography classrooms.

The chapters of this section were written specifically to help guide you through your first instructional experiences and to provide you with a source of reference for your teaching questions in the future. Their aim is to help you make the connection between the advice given in the text and the disciplinary context—in other words, to supply you with geography-specific advice and examples. For each of the eight meetings, this section provides accompanying notes, a series of questions to guide you in your reading of the text, and details of the activities. The facilitator of your sessions may, of course, wish to supplement these readings with others.

This guide focuses on some of the "big picture" issues involved in starting to teach, but you can also learn from other volumes dedicated to preparing new college and university teachers. Some of the best sources on designing significant learning experiences for your classes is the classic *Teaching Tips* (McKeachie and Svinicki 2005); *Creating Significant Learning Experiences* (Fink 2003); *Tools for Teaching* (Davis 1993); *Teaching for Quality Learning at University* (Biggs 1999); and *Classroom Assessment Techniques* (Angelo and Cross 1993). Also of great value are Walvoord and Anderson (1998), Wiggins (1998), and Wiggins and McTighe (2005). For geographers, chapters in *Aspiring Academics* (Solem, Foote, and Monk 2009) provide excellent overviews of course design, active pedagogy, ethical issues in teaching, teaching for diversity and inclusion, and other topics. Background in learning theory is also useful in getting started, and a number of books have been published for college teachers in the last decade or so that provide excellent overviews (Bransford, Brown, and Cocking 1999; Zull 2002; Svinicki 2004).

A note about terminology: in this section, the weekly or biweekly gatherings are called "meetings," consistent with a communal style of interaction, whereas the periods in which you teach are called "classes." The "facilitator" is the person—perhaps a faculty member or senior graduate student—in your department responsible for scheduling the meetings. If your meetings are informal, perhaps the facilitator position will rotate to a new leader each week. Similarly, new geography graduates or early career faculty are referred to as "teachers," whereas the people in the classes you teach are referred to as "students."

Final words of introduction—remember, this section was written for *you* as a new college teacher. It should be considered a "work in progress" that is flexible enough to suit the future needs of the department and the graduates. Therefore, if you think these resources can be improved, it is your job to suggest changes that the authors might make in future editions, so that those coming after you benefit from your insight. The best way to share your ideas and experiences with the chapters in this section is in the discussion forums available on the web site for *Teaching College Geography* (www.pearsonhighered.com/AAG/).

Sample meeting sequence for Teaching College Geography

Before starting the academic semester/quarter: Meetings 1 and 2 (read Chapters 1 and 2)
The week after the first class: Meeting 3 (read Chapter 3)
The week prior to midterms: Meeting 4 (read Chapter 4)
The week after midterms: Meeting 5 (read Chapter 5)
The weeks leading up to final exams: Meetings 6 and 7 (read Chapters 6 and 7)
Around the end of the semester/quarter: Meeting 8 (read Chapter 8)

References

Angelo, T. A., and K. P. Cross. 1993. *Classroom assessment techniques: A handbook for college teachers*. San Francisco, CA: Jossey-Bass.

Biggs, J. 1999. *Teaching for quality learning at university: What the student does*. Buckingham: Open University Press.

Bransford, J. D., A. L. Brown, and R. R. Cocking, eds. 1999. *How people learn: Brain, mind, experience, and school*. Washington, DC: National Academy Press.

Curzan, A., and L. Damour. 2000. *First day to final grade: A graduate student's guide to teaching*. Ann Arbor, MI: University of Michigan Press.

Davis, B. G. 1993. *Tools for teaching*. San Francisco, CA: Jossey-Bass.

Fink, L. D. 2003. *Creating significant learning experiences: An integrated approach to designing college courses*. San Francisco, CA: Jossey-Bass.

Lambert, L. M., S. L. Tice, and P. H. Featherstone, eds. 1996. *University teaching: A guide for graduate students*. Syracuse, NY: Syracuse University Press.

McKeachie, W. J., and M. Svinicki. 2005. *McKeachie's teaching tips: Strategies, research, and theory for college and university teachers*, 12th ed. Boston, MA: Houghton Mifflin Co.

Solem, M. N., K. E. Foote, and J. J. Monk, eds. 2009. *Aspiring academics: A resource book for graduate students and early career faculty*. Upper Saddle River, NJ: Pearson Education, Inc.

Svinicki, M. D. 2004. *Learning and motivation in the postsecondary classroom*. Bolton, MA: Anker Publishing.

Walvoord, B. E., and V. J. Anderson. 1998. *Effective grading: A tool for learning and assessment*. San Francisco, CA: Jossey-Bass.

Wiggins, G. 1998. *Educative assessment: Designing assessments to inform and improve student performance*. San Francisco, CA: Jossey-Bass.

Wiggins, G., and J. McTighe. 2005. *Understanding by design*, 2nd ed. Upper Saddle River, NJ: Pearson Education, Inc.

Zull, J. E. 2002. *The art of changing the brain: Enriching the practice of teaching by exploring the biology of learning*. Sterling, VA: Stylus.

CHAPTER 1

Beginnings

Teresa Dawson

CREATING COMMUNITY

Do you feel like an impostor in the classroom? Are you secretly terrified that your students will quickly discover how little you know? Guess what—you are not alone. One of the best ways of allaying such fears is to share them with others and, in doing so, discover that you have a whole room full of empathy and support. This first meeting sets the stage for graduate students and early career faculty to develop a community that believes problem solving should be mutual and success should be shared (remember the terminology being used in this section: "meeting" refers to the weekly or biweekly gathering between you and your colleagues, mentors, and others with whom you are studying and discussing the chapters and activities).

Questions to Think About Before the Meeting

- What assumptions do you think new teachers make about the teaching process and about students? How might these assumptions lead to problems that arise?
- There is often a "distance" between the teacher and students created by different sets of expectations on the part of each. Coming from different countries is just one possible source of such distance; gender and age are two others. Consider the factors that could potentially distance *you* from your students. Are there times and situations where you would maintain a different distance than in others?
- Think back over your life. Can you think of one clearly exceptional experience you had as a student or learner in any setting, with or without a teacher, inside or outside a classroom? Perhaps this was a moment

when a concept "clicked" or a difficult idea finally made sense. What characteristics of this experience made it so special and why?

- Many teachers spend a restless night or two in advance of their first class, conjuring images of stressful scenarios playing out in the classroom. What "horror story" do you envisage telling your own graduate students in twenty years?

PREPARING FOR YOUR NEW ROLE AS A COLLEGE GEOGRAPHY TEACHER

Perhaps you will someday receive a teaching award or other similar form of acknowledgment attesting to your enthusiasm and skill in teaching. What happens in between now and then is largely a function of you—your needs, aims, and wants. You may never have taught before, or you may be an experienced hand. You may think you know a lot about teaching because you have been on the other side of the red pen, or you may be terrified when you think of the new responsibility thrust upon you. You may even not yet have a teaching assignment, but may hope for one in the future.

Whatever your situation, the aim of the eight chapters and ten activities in this section is to provide as much help as possible in preparing you for the seen and unseen that lie ahead. I hope that these chapters and activities will fire a passion for teaching in you, encouraging you to go to discussions and lectures on campus and to become an award-winning teacher. But, let us be realistic, teaching does not excite everyone to the same degree; if you acquire sufficient confidence to get through the first class without dissolving into a nervous heap, this guide will have gone a long way toward serving its purpose.

Within a typical geography department, teaching roles and responsibilities vary considerably. Graders, as the title implies, usually grade papers and exams for professors and have relatively little contact with students. Teaching assistants (TAs) usually have autonomous responsibility for laboratory and discussion or recitation sections, while in American universities, lecturers, instructors, and professors usually lead whole courses (lecturers and instructors are usually either advanced graduate students or people already holding advanced degrees but not a tenure-track appointment). Labs and discussion or recitation sections, as they are variously known, are small groups taught by the TA alone to supplement and complement the lectures given by the professor or instructor. Again, responsibility within the lab and recitation situation varies greatly according to the professor teaching the course. Some professors will provide TAs lecture notes to present for lab or recitation; others will expect TAs to produce their own.

The length of lab and recitation sessions varies by subdiscipline, as does the overall emphasis of the sessions. Lab sessions for large introductory physical geography may use lab manuals, which dictate quite closely what material should be covered, and do not require the TA to talk to the class for very long out of what is often a two-hour period. Introductory human geography sections tend to be shorter but usually require the TA to conduct the class for the full fifty minutes, though usually allowing TAs more latitude in developing teaching and lecturing materials. Introductory courses in geographic information science (GIScience) may be longer or more frequent, and are sometimes quite highly structured with TAs not lecturing, but rather providing one-on-one help to students in a computer laboratory. When it comes to grading, some professors give specific grading keys, others ask TAs to construct their own.

The degree of teaching autonomy given to graduate students and early career faculty varies considerably from department to department and from institution to institution. Early career faculty taking new positions in departments with graduate programs are often given almost complete autonomy to create and teach courses. However, this autonomy is always tempered by the need to align the course with others in the curriculum, particularly when introductory courses serve as prerequisites for advanced courses. In some institutions such as two-year colleges and some undergraduate geography programs, course content, assignments, and assessments are set in general or even in detail by committee, or by the institution itself. And, in almost all cases, new faculty are often helped by the syllabi, notes, and assignments used by previous instructors of a course.

Similarly, many geography departments feel confident in allowing new TAs a degree of autonomous teaching from the start. If this concerns you, do not worry. Most professors understand your concern and do not require their TAs to lecture and lead discussions immediately from the first day without careful guidance and help. This gives you a small breathing space in which to prepare mentally and physically for your first class. If you have dealt with these issues before and think it is easy, take another look. There may be something you have missed. You will also be invaluable in helping those who are new to find their footing.

One key component of preparation for the first week of class is psychological. As we step into the role of teacher, we are moving also from the status of being a "novice" to that of an "expert." However much you might resist this label, it is one your students will tend to apply to you. Adjusting to your new role can be difficult at a time when you are probably feeling least secure in your ability to fulfill the demands placed upon you. This is a crucial time to boost your confidence. Try talking to others who are (or have been) in the same position. You will soon realize that you have a great deal to offer, and getting into the habit of reflecting on your progress as a teacher will continue to help you.

A very effective aid to self-reflection about teaching involves keeping a journal. In it, you can record such things as tips from experienced teachers, feedback from students, your reactions to a particular class, inspirational moments, issues you would like to discuss with others, and any other ideas that may occur to you. Such a journal will document your growth as a teacher and will be an ongoing source of ideas for the future.

A FEW SOOTHING WORDS

In case you still find all of this very daunting, here are a few thoughts from a TA in Introductory Human Geography, reflecting on his first teaching experience. You should find them encouraging.

> Although the first class did not go exactly as I had hoped, I feel pretty good about the overall results. Among the positives, probably the most important was confidence building. I went into the class a little nervous; not so much because I was going to be speaking in front of a group, but because I was not sure that I was going to be able to get the points I felt were important across to the students in an understandable, logical, and coherent manner. In retrospect, I believe that I was able to achieve this, and thus my confidence has grown.
>
> Considering it was my first time in front of a university class, I am relatively pleased with the results for the first meeting. Although I disliked the uneasy feelings and nervousness, these have quickly passed, and I realize that these are things that most everyone feels. My confidence has been boosted tremendously, and I head into the upcoming weeks with the feeling that the upcoming classes will be enjoyable.

Besides being mentally prepared for next week, the other key component of your preparation will involve making detailed decisions about what you are actually going to do in class. Therefore, in the second meeting, you will have a hands-on activity that helps you construct your first session plan.

But first, remember those questions posed at the beginning of this chapter? Spend the remaining time in this first meeting discussing those questions among the members of your group. There are no right or wrong answers—the questions are simply designed to encourage your thinking about that first class, just around the corner!

CHAPTER 2

Rolling Up Your Sleeves

Teresa Dawson

The first day of class sets the tone for the rest of the semester. A well-run class is like a savings account. Investments you make early on will be paid back with interest later in the course. Getting to know your students and showing that you are interested in their education goes a long way toward establishing a comfortable classroom climate. Similarly, careful preparation for the first day will help define the form the class will take. Having a thorough plan will also make you less nervous on your first day teaching. Therefore, the purpose of this meeting is to help you plan the nuts and bolts of the first class period.

Things to Do Before the Meeting

- If you are the teacher:
  - have your syllabus ready to distribute either online or in hardcopy;
  - develop a good activity or presentation that overviews what students will gain from taking your course;
  - establish the ground rules for the class and communicate these in positive terms to the class, that is, for example, why keeping up with the readings will help with grades; why class participation is important to the course; how to take notes effectively from the visuals, etc.;
  - if teaching assistants (TAs) are assigned to your course, meet with them to specify TA assignments and duties and set up a regular time for weekly meetings.

- If you are a TA, meet with the professor in charge of the class to:
  - establish the ground rules for the class;
  - set up a regular time when you can discuss the course's progress;
  - obtain your specific TA assignment duties for next week and bring these with you to the meeting. (Please make every effort to do this because it will help you greatly in the meeting. However, *do not panic* if your professor is unavailable and you cannot get next week's instructions.)

Questions to Think About in Advance

- Different teachers establish different ground rules for their classes depending on what they personally find acceptable; other ground rules are set forth as an "honor code" by departments or universities. What behaviors are you prepared to tolerate? For what behaviors do you need to set policies on the first day?
- Describe a class you enjoyed as a student. Is this the kind of class in which you would be comfortable as a teacher? Why or why not?
- Given your teaching assignment for next week, what is the main point that you want to convey to the students?
- While it is likely that you will have to deal with classroom problems *after* they have occurred, perhaps even more important to managing the classroom environment is *preventing* problems from occurring in the first place. What can you start doing on the very first day to avoid future problems?

PREPARING TO TEACH AND CLASSROOM MANAGEMENT

Careful preparation is a key to successful classroom management because it enables you to anticipate and prevent many potential pitfalls before they occur, as well as to use unexpected situations to explore issues in more detail and greater depth. Checking out the logistical and physical classroom situation, preparing the details of what is to be done in class, and paying attention to those first crucial minutes are some of the most important preparations you will be glad you made.

Pre-Class Details

Whereas getting to know your students will be an ongoing and at times unpredictable project, the classroom "space" is something you can know almost immediately. Before your first class, visit the room where the class will meet. Familiarizing yourself with the classroom will help you gain control over at least part of an unfamiliar situation.

Within geography this generic advice should be extended to include the various forms of audiovisual (AV) equipment that is so important to teaching in the discipline. Even the most experienced teacher can be thrown off track by the nonappearance of a video booked to fill a full fifty-minute class (yes, it happens). We will assume that such an event will not happen on the first day (no one could be that unlucky, right . . .?). But as the best actors will tell you, it is still a very good idea to check your theatrical context and props in advance. The following questions can help you take stock ahead of time of AV needs for a class.

- Do you need maps for your class (we are geographers after all)? Where are the maps? Take a minute to see how they are cataloged and what is available. Always refile maps immediately after use as a courtesy to others. Also look for online resources—they may be more effective.
- How does the office photocopier work?
- Do you need to make handouts? Do they get posted on the web?
- What course management system does your institution use? Do you and/or your students need password access? Do you or they need to be loaded into a database for this to happen? Who helps with that?
- What idiosyncrasies does your classroom AV or lab technical equipment have? Is there a spare bulb for the data projector, and how do you access it? Do you need keys for the AV cupboard?
- When you check out the classroom, are there any "blind" spots for students? How large does the text need to be on your overhead or PowerPoint slides? What level of lighting allows students to see slides as well as to interact with each other? Do you need to take your own laptop or is one provided? How does your presentation (if applicable) get loaded for use?
- Is a video or DVD required for this or future classes? Does it have to be ordered/made into a clip for playing?

And finally, since no plan is perfect, when using any type of AV equipment, *always make sure you have a contingency plan.*

The Session Plan and Class Outline

Now that the scene is set and you know the limitations of the stage, you need to begin to write the play. The session plan is a detailed account of what you plan to do—even to the extent of writing your lines and choreographing your moves if you feel so inclined. The process of writing a plan forces you to clarify for yourself what you are trying to accomplish and how best to do so. Consequently, spend some time on creating your first session plan. The first one may be challenging but do not give up. To help yourself, try using the outline notes and example in Tables 2.1 and 2.2. (This is what Activity 2.1 asks you to do.) In particular, do not skip the rationale for the class. Everything else really flows from it. Ask yourself, "why should my students learn what is to be covered today, and how can I best connect it to their existing knowledge and experience?" As you get used to thinking in this way, the plans for subsequent weeks will become much easier.

TABLE 2.1 Session plan template

Course Week Date

a. *Setup and announcements*
   Changes in schedule, grading to hand back, assignment due dates, name cards if used, handouts, etc.
b. *Student questions and issues*
c. *Outline*
   Topic, goal, and rationale for class
   Introduction (including links to lecture/assignments)
   Key points in logical order including:
      Concepts/definitions
      Examples
      Concluding summary/reminders
d. *Things to think about in advance*
   Props required (maps, handouts, overheads, movie, AV equipment, etc.)
   Examples you will use
   How to check student understanding
   Where student participation is required and how to elicit participation
e. *Afterward*
   Make a note of each student's performance for participation purposes.
   Write exam/test questions for the class.
   Note any improvements that can be made to outline/session plan.

TABLE 2.2 Example of a session plan for a class on global warming (an edited version of how a teaching assistant for an environmental geography course organized a viable class based on the professor's instructions: "Show the video 'Hot Enough for You?' and then have students discuss the global warming issues, which are the focus of the video")

Course: **Geography 10** Week: **1** Date: 14 **September 2007**

a. *Setup and announcements*
   - Introduce myself
   - Ask students to make name cards
   - Administer "getting acquainted" questionnaire
   - Several details about labs:
     Attendance and participation
     Assignment 1
     Lab Manual
     Calculator
   - Handout of global warming questions
b. *Student questions and issues*

c. *Outline*

Topic, goal, and rationale for class: To discuss the human causes of global warming. We will use the video and learn to read graphs of key variables to understand the complex issues involved and to talk about them in an informed way.

Introduction: Map showing world's population distribution near coastlines and potential for inundation. Importance of global warming for producing rising sea levels. Importance of human contribution to the warming.

Key points in logical order:

1. Refer them to questions on the handout.
2. Show the video (20 mins.).
3. Discuss transparencies (and allow time for written responses afterward).

***Transparency 1* ("Contributions to global warming"):**

- What gases are responsible for causing the greenhouse effect and global warming?
- What kinds of human activities have contributed to global warming?

***Transparency 2* ("Total energy use by fuel type, 1989"):**

- Which are the OECD countries?

***Transparency 3* ("Trends in global energy consumption"):**

- Identify typical levels of commercial energy consumption for both industrialized and developing countries.
- Identify typical levels of per capita commercial energy consumption for both industrialized and developing countries.

Concluding summary/reminders: Emphasize the importance of the human component of global warming.

d. *Things to think about in advance*

Props required: Maps of population distribution, handouts, overheads, movie, AV equipment, questionnaire, name cards, 3 thick markers.

Examples: How much energy students use in a day for showers, study, keeping beer cold, video games, etc. vs. energy used elsewhere in the world.

How to check student understanding: Response to the figures and tables. Collect and read student answers to the questions on the handout. Respond to their answers/any difficulties next class.

Where student participation is required and how to elicit participation: Discussion of questions on handout and issues related to human causes of change.

e. *Afterward*

Note of students' performance for participation purposes.

Exam/test questions from the class:

1. Which group of countries uses the most oil?
2. Write one sentence that describes a difference between the use of energy in non-industrialized and industrialized countries.
3. Of the ways in which human activity contributes to global warming, which do you feel is the most important and why?

Note any improvements that can be made to outline/session plan.

- Shorten time on video and use more time for discussion.

Once you have written the session plan, you can provide students with an outline to emphasize its main points. There is a threefold advantage to using an outline. First, it helps students to see where you are going. Second, it keeps you on track. Third, it can act as a lifeline. If you are nervous and forget what you are going to say, you need only look to your outline for a moment to get back on track. The students will not even notice or will think you are generously giving them a few seconds to digest your last point. Keeping a file of session plans and class outlines will prove an invaluable resource if you ever have to teach the course again and, if you are of a generous nature, will also provide tremendous comfort and support to other teachers who might teach this course in future semesters.

Two of the greatest concerns that teachers have about teaching the first class involve how to check for students' understanding and how to get students to participate in meaningful discussion. The two are not unrelated. If effective explanations have been given, students will be better equipped to voice informed opinions. Getting students to participate—for example, by verbalizing how they might apply a general concept to their own experience—is an excellent way of checking for understanding. Because student participation and checks for understanding are central to the success of many labs/recitations in geography, it is important to think in advance how they will be built into your session plan. Many teachers report that the most useful tool for eliciting participation from students is knowing who they are, which means knowing their names and knowing where they come from in the broadest sense. You might also try a one-minute paper at the end of class to see if the students' idea of the main point of the class corresponds with yours (be prepared for some surprising results).

Whatever you decide to do to maintain student participation and understanding at high levels, you will want to set the precedent for your approach on the very first day of class. In this respect the first few minutes can be key.

The First Few Minutes

The first few minutes of the first class are perhaps the most intimidating, yet they also provide a great opportunity for both you and your students to get to know each other and the class. Set the tone by being there well in advance and chatting with students informally as they come in. It is also probably wise to check that everyone in the room is in the right class. Then start on time.

- *Introduction and details about you*—introduce yourself. Give some key details students need to know about you and display them on the board (your name, office hours, office phone number, and address). If you are a foreign-born, international teacher, this also serves as "tuning-in time"—a period where the students can get used to your accent before they really have to start concentrating. Tuning in is much easier in a low-pressure situation like writing down office hours than it is if you launch immediately into detailed material full of new terminology. Having volunteered information about yourself, you then need to get acquainted with your students.

- *Name cards*—you might try using name cards for the first few weeks until you learn everyone's name. Giving the cards out and collecting them helps to connect faces and names. Also, if you are required to take attendance, this is a good, indirect way of finding out who is there (according to whether their card was used or not).
- *The "getting acquainted" questionnaire*—in the spirit of getting to know your students, giving them a "getting acquainted" questionnaire is a good idea. Either use the sample questionnaire in Table 2.3 or construct your own. Such a questionnaire will demonstrate that you are interested in your students as people, help you to know their names, and inform you of their aspirations for the course, thereby enabling you to tailor the examples you use to their experience even if you cannot adjust the content matter. This information will be important when you are subsequently trying to obtain feedback on your performance at a later date in terms of your ability to meet student expectations. You could also include such items as "What are your major interests outside of school?" and "What is your major?" One teacher I worked with picked three questions *she* wanted to ask and then allowed the students in the lab to choose two additional questions *they* wanted to answer. In this way, she indicated to the class that she was happy to let them set some of the agenda for the

TABLE 2.3 Sample "getting acquainted" questionnaire

In a large introductory class, is it very easy to "become lost" in the system. Often an instructor will only know a name and a student number. I want to avoid this. One of the purposes of the recitation section is to allow me to get to know you as an individual. In this way, I can help you with your work in geography and with any problems that might occur. I want to be able to tie together faces with names *and* with personalities. It helps me, therefore, to know a little of your background.

Would you please spend a few minutes answering some questions? If you feel that these questions are inappropriate, please do not feel compelled to answer them. Thank you.

Your name:

Your major field of study:

What are some of your interests?

Where is your home?

Please list any previous geography courses:

Reason for taking the course:

If possible, could you list the sort of things that you hope to get out of this course?

class within the limits she had prescribed. Finally, you should try to acknowledge comments that students have written on their questionnaires in the very next class.

- ***Details about the course and transition to the topic for the day***—after completing these preliminary introductions, you can explain key course details you feel they need to know—the course structure, the syllabus, requirements, grading policies, assignments, etc.—and ask if they have any questions about them. Your professor may have planned that you devote the entire first period to such a discussion. If not, you can now move on to the topic for the day as outlined on your session plan. And you, with the first few minutes over, will be standing on firm ground.

Activity 2.1: Creating Your First Session Plan

Use your specific teaching assignment for next week, the sample session outline, notes, and example provided (Tables 2.1 and 2.2) to put together your first class. The session outline was adapted from the one I have used generically. A group of geography teachers ran trials to establish the most important features of a plan for an average session in geography, and the sample given is the result of their collaborative effort. You should feel free to modify it to your needs, however (no two geographers will have exactly the same needs).

Notes on Completing the Session Plan

Setup and announcements—includes changes in schedule, homework to hand back, assignment due dates, name cards if used, handouts, etc. Get the students used to knowing that you will give important announcements at the start of class and do it in a consistent fashion so that if they are late they know whether to look to the upper-left-hand corner of the board or for a handout, etc., for the information. This approach should help them realize the importance of being on time.

Student questions and issues—use this time to clarify any points from the last class session that may not have been clear or to ascertain whether they have any other questions that need answering. Use your judgment here. If the question is likely to affect the whole class and can be answered quickly, or if today's class depends on the students understanding the point, then respond. If, however, a student has a question that needs answering on an individual basis (*e.g.*, a question about a grade on a paper), this is better handled in office hours or after class.

Outline—the following are some key elements to organize as you plan the sequence of events in the class.

- ***Topic, goal, and rationale for class***—give the topic for the day. State clearly what is to be achieved and why. This will help both you and your students to focus on why the class will be helpful. Include any links to main lectures/assignments, if appropriate.
- ***Introduction***—try to establish a connection between the topic and student experience by using a focusing example, anecdote, or problem.
- ***Key points***—list in logical order, including the key concepts/definitions that need to be put across and examples that can be used for illustration.
- ***Concluding summary and reminders***—refer back to the goals and show how they have been achieved. Remind students about commitments for next week such as reading, etc. Tie next week's class to what you have just covered.

Things to think about in advance—make a list of things you need to take to class with you and organize them well in advance.

Afterward—it is helpful to think about possible exam questions based on the class as a way of considering whether you fulfilled the class objectives. Reflection on student participation and on your own performance is best done immediately after class so that it is fresh in your mind. This will also help you in teaching the next section.

CHAPTER 3

Into the Lion's Den

Teresa Dawson

In this chapter you will put into practice the ideas that you developed in the previous chapter, in particular, you will try out your session plan. These preparations should make your first day in the classroom as smooth as possible and will ease your transition to teacher in action. This week's meeting is designed to take place after everyone has conducted their first class and had a chance to reflect upon it. The meeting will give you the opportunity to reflect upon the positive and negative aspects of your first classroom experience and to congratulate each other on your success.

Things to Do Before the Meeting

- Solicit ideas about lecturing or conducting a class discussion from your mentor, advisor, other professors in your department, or from staff in the "teaching and learning center" on your campus (if one is available).
- Prepare well and in conjunction with feedback from others who have taught the course before. If you are serving as a teaching assistant (TA), you will want to compare your plan with those prepared by other TAs assigned the same course, or, if you are the only TA assigned to the course, with those of TAs who have taught the class before or with the professor leading the course. Use the session plan, the getting acquainted questionnaire, ideas from our last meeting, and the readings to help you.

Questions to Think About in Advance of the Meeting

- Choose one concept that your teaching assignment requires you to convey to your class and think of *three* different ways to explain it.
- Both "lecture" and "discussion" classes involve a high degree of student participation. Using your concept from the previous question, how will you get *your* students to participate in the learning process? Chapter 10 "Looking Beyond the Lecture" in this book is a good source of useful ideas.
- What principles of good teaching appear to apply to both lecture and discussion teaching?
- Suggest one compelling idea from Chapter 10 or your hallway conversations this week that you think would be particularly appropriate to enhancing your interactions with your students.
- Immediately after you teach, but before this meeting, reflect on the following:
  1. Which aspects of the first class did you like?
  2. Which aspects would you like to improve on?

TEACHING METHODS

So you survived the first class? Were the students less terrifying than you thought? Did you remember to write possible quiz questions straight after class when they were fresh in your mind? Did you get a discussion going? Did you run out of time?

Having reflected on these questions, what additional steps do you need to take to make your performance even more effective? As stated last time with regard to keeping a journal, self-reflection is an important part of improving your teaching. With this in mind, Activity 3.1 is designed to help you reflect effectively on your first class—what went well and what could use some improvement.

As you become more reflective about your teaching, you will probably need to be less concerned with your own personal survival and correspondingly will be more open to the needs of your audience. Although it may not be immediately obvious, individuals often learn (and teach) best in quite different ways (although often we tend to teach best the way we learn best). Think about how you prefer to study. Do you like to have concepts explained to you verbally? Do you prefer to read the original work explaining a discovery? Or do you have to build a physical model incorporating the concept in order to understand? In other words, if you prefer to lecture to your students, you may not be reaching those who learn best by discussion. Certainly no one would wish you constantly to teach in a mode you find uncomfortable. However you might want to consider varying your teaching method and your explanations to reach a broader audience. How might you alter your approach in different cultural contexts or facilitate the learning of students with disabilities?

EFFECTIVE EXPLANATIONS IN GEOGRAPHY

For whatever reason (and there are many debates over this) some of us learn better in some ways than others—for example by listening to a lecture, by reading a textbook, by discussing ideas with friends, or by building a working model. There is even some evidence that people with certain preferred learning styles may be attracted to certain disciplines (Healey et al. 2005). Therefore, it would appear that one of the most effective ways to begin serving the needs of all the students in your class would be simply to ask yourself how they learn best on an individual basis, and then to adjust your approach accordingly.

However, geography teachers at universities and colleges face a particularly challenging situation when it comes to accommodating the learning styles of students in their classes. This is because our discipline covers subject matter that stretches from the humanities and social sciences through the physical sciences and even into the professional schools. In addition, departments often run a number of large introductory courses that serve to fulfill the general education and diversity requirements for many nongeography majors. The result is that few generalizations can be made about the students you are likely to encounter.

Because you are likely to face—particularly in the large introductory classes—a wide range of nonmajors from varied backgrounds, you need to be as flexible as possible (within reason) in varying the explanations and examples that you use in class. It helps to include visual, verbal, and practical elements, and to relate the concept to the students' own life experiences as much as possible.

When teaching a new concept, it is also worth considering the following steps of concept teaching recommended by, among others, Savion and Middendorf (1994):

1. Introduce the concept in a vivid, familiar example; this helps tie the new concept to old knowledge. In class, describe something college students are familiar with, something that they would have on their minds, or tell a story with lots of details or show a video clip to provide them with a detailed example.
2. Introduce an unfamiliar context for the concept, one that is unlikely to be close to the students' everyday experiences.
3. Once you present the concept in simple and theoretical applications, a definition may enhance the students' comprehension by smoothing the edges of the informal representation and thereby providing a rule for correct production.

4. Students should practice thinking in terms of the new concept to really make it part of their knowledge. Have them do something with the concept, such as teaching it to someone else, developing their own examples, or summarizing it in their own words.

The following is an example of how you might put these principles into practice.

Example: Teaching the Concept of Mental Maps

Suppose you need to explain the concept of a mental map in a section of introductory human geography. You might begin by asking students to draw a sketch map of the college campus. Then have them swap with a neighbor and let each suggest what they can about the other based on their map. For example, perhaps it is obvious they have not been in town long because their map is not very detailed, or they may like to enjoy a very active social life because they know where all the bars are, or it is likely they do not have a car because their map focuses only on the immediate area, and so forth.

Next you could show your students some published pictures of mental maps (examples of which can be found in most human geography textbooks or on the Internet), and ask them what they might deduce about the authors of these maps. For example, "Can you tell how old the authors are? What is their likely level of mobility? What cultural factors or symbols are important to them?"

Once a certain level of understanding has been achieved, you could then ask students how they would define the term "mental map." You could then write this information on the board or on an interactive slide on the course page for students to record, and at the same time verbalize the class consensus aloud. For example, the consensus might be that "a mental map is an image or a map of a place that you have in your head, the form of which is likely to be affected by your experience and beliefs."

Finally, to check for understanding, you might ask the students for examples of contexts in which geographers would employ mental maps, such as for planning city space for children or the elderly, or designing a campus map for new international students at your university.

THE ART OF ASKING QUESTIONS

In addition to giving effective explanations, another aspect of your teaching that often needs special consideration involves the art of asking questions. The types of questions you ask students will in turn determine the fullness and depth of the response they give you and the type of learning that takes place. If students are not responding to your questions in the way that you would hope, consider what kinds of questions you are asking them. Not all questions are created equally as shown by the following examples of six types of questions adapted for geography.

Type 1

a. Does everyone remember what Central Place Theory is?
b. Do you all see the difference between the Gall-Peters and Mercator projections?

These are essentially *rhetorical questions,* questions that will provide you with little or no response and will not enable you to check for understanding in your students.

Type 2

a. Now we have this table. What's going on? Which countries do we mainly have figures for? What's happening in the rest of the world? How do you calculate the percentage change? How is this different from the table we saw last time? What's the data source?

Here we have the *"how many questions can you squeeze into one question" question,* questions that will confuse your students and will often silence them because they are unsure which questions they should attempt to answer.

Type 3

a. What do you think happened in 1973 to cause this change? Was it the oil crises brought about by OPEC?
b. What happens when we add the sum of the rows? Do we get skewed results?

These examples represent *leading questions,* questions that give students a considerable hint as to the answer and make it seem unnecessary to respond given the "obvious" nature of what is involved. These questions will not encourage students to think critically or to give very full answers. In addition, students will quickly learn that if they only wait long enough, you will answer the question for them. However, these questions can sometimes be used with humor when there is an impasse and then followed up with further explanation.

Type 4

a. I think calculating the lapse rate here is pretty straightforward. Any questions?
b. Who can reword her answer the way you think I would say it?
c. Anyone so confident in their answer that they want to write it on the board?

This group represents *put-down questions,* questions that belittle students. Such questions are not recommended. They will kill participation now and in the future and will lose you the respect of your students.

Type 5

a. Do you notice this business of noise?
b. What do you think is going on?

Such formulations are *open-ended but vague questions,* questions where it is not entirely clear what you are getting at or how a student might formulate a response. The sentiments behind such questions are fine—the question itself just needs clarifying. See if you can reword the examples given.

Type 6

a. How might the political map of Europe look different today if the U.S. had supported Germany in WWII?
b. What implications do you see for the urban and social geography of New York City resulting from the recently proposed crime bill?
c. What evidence would you look for to determine whether global warming was caused primarily by human activity or by other factors?
d. What is the most effective way of representing the data you have?

This final type represents *thought-provoking or response-provoking questions,* clear questions designed to elicit full and critical responses from students and encourage them to ask further questions of each other or you.

Of course, the "put-down" should be avoided, but most of us use versions of the others at some time or other. The key is to be aware of your purpose in asking the question, and then to ask the kind of question that best accomplishes that purpose. For example, a rhetorical question such as "Does everyone remember Central Place Theory?" can serve as a useful transition, but it most likely will not encourage thoughtful discussion. If you want students to think critically and to verbalize their problem-solving process, try to increase your use of open-ended, thought-provoking questions such as "Do you think Christaller's model of Central Place Theory would apply in Pennsylvania in the twenty-first century?" This may improve class response in terms of both volume and content.

THE RELATIONSHIP BETWEEN QUESTIONS AND KNOWLEDGE

The type of question you ask, then, will often determine what kind of learning takes place in the classroom. When asked, most teachers say that they would like their questions to promote critical-thinking skills or "higher-order" learning among their students. Such learning refers to the latter categories of Bloom's Taxonomy of Educational Objectives in the Cognitive Domain (Bloom 1956), concepts that are not new, yet still very much relevant to contemporary teaching practice. The following is a brief description of these categories and suggests appropriate questions that may be used to invoke them:

1. ***Knowledge***—the remembering of previously learned material, measured by recall of items from specific facts to complete theories. (*Who . . . ? What . . . ? When . . . ? Where . . . ? How much . . . ? Define . . .*)
2. ***Comprehension***—the ability to grasp the meaning of the material, demonstrated by translating material from one form to another, interpreting material by explaining or summarizing and estimating future trends by predicting consequences or effects. (*State in your own words . . . Give an example . . . Explain . . . Classify . . .*)
3. ***Application***—the ability to use learned material in new and concrete situations, using rules, methods, concepts, principles, laws, and theories. (*Predict what would happen if . . . Judge the effects of . . .*)
4. ***Analysis***—the ability to break down material into its constituent parts so as to understand its organizational structure, involving identification of parts, investigation of relationships between the parts, and recognition of the organizational principles. This requires the understanding of both form and content of the material. (*What persuasive technique is used . . . ? What assumptions . . . ? Make a distinction . . . What inconsistencies . . . ?*)
5. ***Synthesis***—the ability to put parts together to form a new whole, involving the production of a unique communication (theme or speech), a plan of operations (research proposal), or a set of abstract relations (scheme for classifying information). Stress is on creativity through formulation of new patterns or structures. (*How would you test . . . ? Propose an alternative . . . How else would you . . . ? Formulate a theory . . .*)
6. ***Evaluation***—the ability to judge the value of material for a given purpose, basing judgments on definite criteria. The student may be given the criteria or determine them independently. (*Judge . . . Compare . . . Which is more important, better, logical, valid . . . ?*)

Using the examples above, consider how you might reword one of your less productive questions in order to get a different kind of response from your students the next time you try it. These suggestions can also be useful if you need to help write exam questions or quizzes for the course (for more on grading and test construction, see Chapter 4).

Activity 3.1: Reflecting on the First Class

Please hand in the following four items from your first class to the meeting facilitator:

1. ***A copy of what you planned to do or what you were instructed to do by the course instructor:*** This could be one sentence or one page.

2. ***A copy of the session plan you filled out for your class:*** This is probably one sheet. Write your name, class name, and any other relevant class details on the top. You need to find something you are comfortable with using every week. The idea is to keep a file of session plans so that if you teach the same course next year you will already be prepared. Even if you do not, it is still excellent practice to get into for when you are a full professor. In addition, it will be a wonderful resource for those who will teach the course in the future.
3. ***A copy of your outline of the contents of the first class:*** Probably one page of large headings. Again, get into the habit of keeping a file of these.
4. ***A short written statement, "reflections on the first class":*** This should consist of a paragraph each on (1) what you liked, (2) what you disliked, and (3) what you are going to try and do differently. If you are not actually teaching, please write a paragraph each on (1) what you think would be most scary about the first class, (2) how you might practically prepare to reduce your anxiety, and (3) what you think might be enjoyable about the experience (be positive—there must be something).

Remember!—it is important to try to act on these reflections in the very next lab/recitation section you teach so that "first-class hitches" never have a chance to become fossilized as problems.

References

Bloom, B. S., ed. 1956. *Taxonomy of educational objectives, handbook I: The cognitive domain.* New York: David McKay.

Healey, M., P. Kneale, J. Bradbeer, with other members of the INLT Learning Styles and Concepts Group. 2005. Learning styles among geography undergraduates: An international comparison. *Area* 37 (1): 30–42.

Savion, L., and J. Middendorf. 1994. Enhancing concept comprehension and retention. *The National Teaching and Learning Forum* 3 (4): 6–8.

CHAPTER 4

Help with Grading

Teresa Dawson

It is usually around this time in the semester when teachers begin to grade the first batch of assignments and to think about midterm exams. Grading usually gives new teachers more worry than any other aspect of an academic job. Perhaps it is the sudden responsibility or the feeling that grading is not such a precise science as they would have liked to think. In any event, several things can be done to relieve grading tension. Most revolve around good communication among professors, teaching assistants (TAs), and students and thinking carefully about how to align the grading with course goals. In this meeting, you will have the chance to develop satisfactory grading criteria and a key, sometimes also called a "rubric," as well as the opportunity to practice applying them to actual student papers in geography.

This chapter is aimed a bit more at graduate TAs or graders who will actually be scoring student work, rather than at professors who will be writing the exams or assignments. Help with writing exams is offered in the second part of Wiggin's (1998) *Educative Assessment;* Walvoord and Anderson's (1998) *Effective Grading;* and Chapter 3 of Fink's (2003) *Creating Significant Learning Experiences.*

Things to Do Before the Meeting

- If you are serving as a TA, meet with your course professor to establish the criteria for grading.
- Read over a sample of the exercises, assignments, and quizzes that you have.
- If there is more than one TA on the course, set a meeting with them to formulate the grading key.
- Bring a sample of the student exercises, assignments, or quizzes to the meeting. Make sure the student material you bring is anonymous (delete any student details so that they cannot be identified).

Questions to Think About in Advance

- How will you respond to student papers in such a way as to help them learn as well as assess their work?
- From your meeting to formulate the grading key, what problems can you foresee in its implementation? How do you propose to overcome them?

GETTING INVOLVED IN TEST AND ASSIGNMENT CONSTRUCTION

Grading and assessment of student performance is an essential part of learning. According to Gibbs (1999), grading is important because it can:

- capture student attention and effort;
- generate appropriate learning activity and work;
- provide feedback to students;
- develop in students the ability to monitor their own learning standards;
- allocate grades;
- ensure accountability (to show outsiders that standards are satisfactory);
- offer feedback to the teacher on pacing, structure, materials, etc., used in the course.

Noticeable in this list is that allocating grades is only one of many functions of assessment, but it is often the most pressing for us. Furthermore, feedback to a teacher can be of very great value. But Gibbs also notes that, to be effective, assessment must be:

- *timely*, so that students can use it for subsequent learning and work to be submitted;
- *supportive of learning*, so students have clear indications of how to improve their performance;
- *focused on achievement*, not effort;
- *specific to the learning outcomes*;
- *fostering independence*, so that eventually students are capable of assessing their own work;
- *efficient*, for you or your graders.

One of Wiggins's (1998) contributions to the theory of grading and assessment is his notion of "educative assessment," which is forward-looking and provides feedback that will help students learn better in the future. He contrasts this with "auditive assessment," which is backward-looking and records what has been accomplished. Though most of us have probably experienced more auditive than educative assessments in our own educations, Wiggins provides good reasons for focusing instead on forward-looking tests and assignments that help students build on their knowledge to better face future challenges.

Effective grading begins with careful test and assignment construction. That means considering some of the grading issues as the test or assignment is put together. Putting such a test or assignment together in turn relies on having a series of well-thought-out session plans and clear learning goals for the course. No one is suggesting that first-time TAs will write the tests and assignments—that is the professor's job—but most professors will at least give their exams to their TAs ahead of time in order to obtain their comments or invite TAs to write some exam questions or some parts of assignments. Take full advantage of this opportunity. It will save you considerable time later on.

Things to Consider

Begin by asking yourself the following for each exam question or part of an assignment:

1. Can my students be expected to answer this question, problem, or exercise given the material that has been taught?
2. Does it require ways of thinking about the world that are intuitive to geographers but have not been conveyed to students? If it does, is the question broken down so as to model the thinking process required?
3. Is the question or assignment worded in the best possible way so as to avoid ambiguity, etc?
4. Have spelling and grammar mistakes been eliminated?

In all cases, consider which type of thinking you are requiring of your students. You may find it useful to revisit Bloom's Taxonomy and the sample questions provided with each category to help you (Chapter 3). Try to get an appropriate mix of higher- and lower-order thinking during the exam or assignment. Many people start with lower-order questions and build to higher-order ones. Also, consider how you are going to reward the demonstration of different types of knowledge.

Beyond these initial questions, your approach will obviously need to vary, depending on the type of exam or assignments given. Here are a few notes on some of the most common types.

- *Multiple-choice questions*—these are perhaps some of the most difficult questions to write well. Begin by taking "The Test" (Table 4.1) This highlights some of the most common problems to avoid. If you have to construct a multiple-choice exam for yourself, use the university testing service or equivalent (ask at your institution's teaching and learning center) to check for validity and reliability in your questions.

TABLE 4.1 The test

1. Trassig normally occurs when the
   a. dissels frull
   b. lusp chasses the vom
   c. belgo lisks easily
   d. viskal flans, if the viskal is zortil
2. The fribbled breg will snicker best with an
   a. Mors
   b. Ignu
   c. Cerst
   d. Sortar
3. What probable causes are indicated when tristal doss occurs in a compots?
   a. The sabs foped and the doths tinzed
   b. The kredges roted with the rots
   c. Rakogs were not accepted in the sluths
   d. Polats were thonced in the sluth
4. The primary purpose of the cluss in frumpaling is to
   a. remove cluss-prangs
   b. patch tremalls
   c. loosen cloughs
   d. repair plumots
5. Why does the sigla frequently overfesk the trelsum?
   a. All siglas are mellious
   b. Siglas are always votial
   c. The trelsum is usually tarious
   d. No trelsa are directly feskable
6. The snickering function of the Ignu is most effectively performed in connection with which one of the following snicker snacks?
   a. Arazma tol
   b. Fribbled breg
   c. Groshed stantol
   d. Fallied stantol

Notes on the Test

The most common giveaways on multiple-choice tests:

1. The correct answer is ***longer*** and ***more precise*** than the distractors.
2. The use of ***common elements*** (or resemblances) in the stem and the correct answer.
3. Inadvertent cues in ***grammatical construction,*** that is, the use of specific determiners (or absolutes) such as all, none, always, and never.

(continued)

TABLE 4.1 Continued

4. The choices lack ***grammatical consistency*** with the stem.
   - the use of articles
   - verb tense discrepancies
5. A cue to the answer to a later question is given in a ***previous*** item.
6. An alternative strikingly ***different*** from the others.

Based on Milton's (1982) mock examination created in gibberish from Lewis Carroll's "Jabberwocky" designed to highlight common test-construction errors.

- ***Short-answer and essay questions***—write answers to the questions yourself (under timed conditions for an exam). Have another instructor or TA who is not from your course answer the questions or at least verbalize what *they* think you are looking for. Break down the point allocation as specifically as you can and write it on the exam or assignment. If students are to write on the paper, leave space appropriate to the depth and length of response you require. On short-answer exams, consider asking a more straightforward question first in order to relax students and ease them into the exam. Professors may ask you to suggest possible test questions based on what you have covered in lab or recitation. This is where the questions from the end of your session plan will come in handy.
- ***Lab quizzes (See also "short-answer and essay questions" above)***—lab quizzes in physical geography are probably the only types of tests that geography TAs will be asked to write autonomously. This does not mean, however, that you are alone in your endeavors; the professors in charge of such courses generally will be happy to give you feedback on your first quiz construction attempts. Also, use the question bank set up by previous TAs for your class and enlist the help of other TAs in the course.

GRADING STRATEGIES

In addition to the other suggestions given in the readings this week, here are some tips from geography teachers past and present who have lived to tell the grading tale.

- ***Hand out criteria ahead of time***—this will keep you on track when you grade (helping to salve your conscience over objectivity issues) and will also be useful if any students query their grades later. In the case of an exam, be explicit regarding what students will be judged on. In the case of a written assignment, hand out your grading criteria with the assignment. Most institutions now have proprietary grading standards and criteria. You might adapt these descriptors for the particular geography assignment in question and/or visit your local writing center for assistance with other tips and suggestions.

- ***Set up a grading key or rubric***—ideally, if you are working as a TA, the grading key should be provided by the professor. If, however, you are required to produce such a key yourself, it is always wise to clear what you have decided with the professor before you try to implement it, especially if the professor will need to explain it to students afterward. Follow the steps you went through before this week's meeting. Essentially this involves keeping in constant contact with the professor and other TAs on the course. The process is so important that it forms the basis for Activity 4.1.
- ***Consider issues of quality control and multiple sections***—it is very important for TAs in multi-section classes to coordinate their grading keys, both with each other and with the professor of the course. Problems of consistency often can be solved by grading one question across sections rather than within a section and by grading "blind."
- ***Maximize the use of your time with a grading sheet or template***—teachers are often torn between their need to get through a large pile of tests or assignments and their desire to give detailed comments to each student. Also, they often find that, by the time they reach the twentieth paper, they have written the same comment at least eighteen times. Prevent this problem and maximize the use of your time with a grading sheet that can be clipped to each student's paper. If you need to keep detailed records (because students need to show improvements over the course, for example), then these sheets can easily be photocopied for your files. A grading sheet might consist of the following (Table 4.2):
  - ***A summary of the common mistakes everyone made***—before beginning, read a sample of the papers through to get a sense of what the common mistakes/misunderstandings are, and type these up. You can then simply check the relevant points for each student.
  - ***A shortened version of the grading criteria (which you may have given to the students in advance)***—again, you can check what applies and underline what has been missed.
  - ***A space for comments***—this is where you can best use your time to address the particular needs of the individual student. Rather than focusing on all the possible improvements or mistakes a student has made, it is often more effective to highlight the two to four most important changes the student might focus on in the future.
- ***Grading participation***—several teachers grade their students on participation in labs or recitation sections. In some cases this can be a substantial part of the student's overall grade, so it is probably best not to leave it to an arbitrary "oh, I remember she said something

TABLE 4.2 Sample grading key or template

Helpful Hints

1. Give the grading criteria to your students when you give them the assignment. Be as specific as possible. Criteria might fall into different categories—for example, regarding specific content, as well as the general nature of an A, B, or C answer.
2. Once the assignments are in, start by scanning a sample of student papers/answers and identifying common errors you would have to address in many of them.
3. Make your *grading key* by typing these errors into your template (see sample below). Refer to lecture notes or text pages for supporting help. Be sure to relate errors/suggestions to the previously identified criteria. Leave room for specific comments at the bottom, so you can individualize your remarks for each student.
4. Make copies of the *grading key* (several will fit on one page).
5. As you grade, put the student's name on the key and attach a copy to each assignment.
6. Simply check the items students have missed and then make a comment particular to each student at the bottom. A balance of "positive reinforcement" with a "suggestion for improvement" is the most effective approach.
7. For maximum impact, photocopy the *grading keys* for students and keep them. Refer back to them in subsequent assignment grading, so that you can comment if a student has taken your suggestions to heart, or if he or she needs to refer back to what you said last time. In this way, students will be amazed that you can remember from one assignment to the next and will take your comments much more seriously.

Course: Assignment:

Name: Date:

Criteria and common errors in this assignment:

A.
B.
C.

Comments and suggestions for next time:

vaguely relevant in the last class" decision. At the beginning of the semester, establish what system you will use for grading participation. The decision may be up to you, but it is usually some combination of attendance and useful contribution. If you are working with TAs, explain your system to them, ask for input, and make sure the standards you develop are applied consistently across all sections led by TAs.

For further good advice on these issues, Walvoord and Anderson's (1998) *Effective Grading* is particularly useful, especially the chapters on "Establishing Criteria and Standards for Grading," "Making Grading More Time-Efficient," and "Using the Grading Process to Improve Teaching." Walvoord and Johnson introduce a useful idea for developing keys which they call "Primary Trait Analysis." This entails having the teacher: identify the traits that will count in the scoring, build a scale for scoring on each trait, and evaluate student work against these scales.

It is also useful to know that good samples of keys for grading different types of assignments are available, such as those found in Hay's (2006) *Communicating in Geography and the Environmental Sciences* and Kneale's (1999) *Study Skills for Geography Students*.

POST GRADING ADMINISTRATION

Although you may feel tired after your first major grading assignment and would dearly love to put all thoughts of the activity aside, be sure to bring closure to the task by taking good care of a few administrative details first. These are always best handled when they are fresh in your mind.

- ***A note on posting grades***—if you post your students' grades, always make sure your students know exactly when and where their grades will be posted (or you are likely to be flooded by e-mail and phone messages). Also, when you post grades, adhere precisely to your institution's policies and procedures (for example, in most places you may no longer post by social security numbers). These days, online systems, which post grades immediately for students to access through their own private accounts, are quickly eliminating the need to post grades in public.
- ***Keep a record***—always keep systematic records of grades, including those for participation. Students may come back to you up to a year after the course and long after you have forgotten what they did or did not do in lab 3. A detailed record will help you to reconstruct what happened. Also, the more specific the breakdown of points (for an essay, for example), the easier it is to see immediately why you gave that student who is breathing fire in your office a "15/25" on her essay.

- *Where your responsibilities end*—do not take on too much. Ultimately, the responsibility concerning final grades rests with the professor. If you have cleared your grading scheme with them, they should back you up. If you are unable, despite your best efforts, to resolve any issue with a student, refer them to the professor if you are a TA or to your department chair.

Activity 4.1: Constructing and Implementing a Grading Key for Your Class

Use the tests and assignments your students have handed in, the grading criteria, and the section on grading strategies (above) to construct a grading key (or "sheet") for your class. As you work, it will help to ask yourself the following questions:

1. What are the key criteria that establish a student's grade?
2. How are the points to be allocated?
3. Were there any common mistakes that can be summarized for everyone?
4. How much space will I need on the sheet for individual comments?
5. What could I do to help my students do better next time around?

Activity 4.2: The Grading Process

Please hand in to the meeting facilitator the following materials after you have completed your first grading assignment:

1. A copy of your grading key.
2. A short statement concerning the problems you encountered and how you overcame them.
3. An example of one question (not necessarily yours) that did not appear to be successful on an exam, quiz, or assignment that you graded and your suggested revision of that question (please provide both old and new versions).

References

Fink, L. D. 2003. *Creating significant learning experiences: An integrated approach to designing college courses.* San Francisco, CA: Jossey-Bass.

Gibbs, G. 1999. Improving teaching, learning and assessment. *Journal of Geography in Higher Education* 23(2): 147–55.

Hay, I. 2006. *Communicating in geography and the environmental sciences*, 3rd ed. Melbourne: Oxford University Press.

Kneale, P. E. 1999. *Study skills for geography students*: A practical guide. London: Edward Arnold.

Milton, O. 1982. *Will that be on the final?!* Springfield, IL: C.C. Thomas.

Walvoord, B. E., and V. J. Anderson. 1998. *Effective grading: A tool for learning and assessment*. San Francisco, CA: Jossey-Bass.

Wiggins, G. 1998. *Educative assessment: Designing assessments to inform and improve student performance*. San Francisco, CA: Jossey-Bass.

CHAPTER 5

Constructing a Feedback Questionnaire

Teresa Dawson

So here we are at Chapter 5 and the fifth meeting. You have (probably) found (and survived) the classroom, dealt with the responsibility of grading something, and even begun to relax a little and enjoy yourself. So perhaps it is time to take stock, to move from "mere survival" to "better effectiveness." Are those silences after your explanations of the principles behind the Mercator projection really the result of awe at the eloquence and beauty of your presentation, or are they despairing incomprehension? Why do those students at the back always seem to be squinting when they look at you? Do they have problems with their eyesight, or could the fact that you write the overheads five minutes before class with a yellow pen be the reason? The aim of this meeting is to increase your teaching effectiveness by setting the stage for the giving and receiving of constant feedback. Such feedback will enable you to more effectively build on your strengths and identify weaknesses that need to be addressed. It will also facilitate the development and strengthening of your teaching philosophy. While your philosophy of, and approach to, teaching are a personal reflection of yourself, there are certain goals and improvements to which we would all aspire. One such goal is to treat all students equally. Therefore, an important component of the next several meetings is to consider how you might best adjust what you do in order to accommodate, as much as possible, the needs of all the students who might be in your classes.

For the next few weeks, you will be focusing on becoming a better teacher. You really only improve your teaching by first obtaining some form of assessment of your current performance. As this week's readings suggest, assessment of your teaching can be achieved in several ways (classroom observation, videotaping teachers in action, etc.). The approach we will take here is to construct a feedback questionnaire and administer it in your lecture, lab, or recitation. Armed with the results of such a questionnaire, you can then follow it up by judging what adjustments could usefully be made. With this in mind, the objective of this week's meeting is to develop your own personal feedback questionnaire that can be given to your students. By doing this now, you will be tackling problems before they become habits and while there is still plenty of time to make appropriate changes before the end of the semester. Join in the discussion even if you are not teaching this semester; you will have a chance to obtain feedback of your own in the future.

Things to Do Before the Meeting

- Look at the sample feedback questionnaires created specifically by geographers (Tables 5.1 and 5.2). Also, check out the question bank of other alternatives generated by geography teachers in Table 5.3.
- Retrieve the original "getting acquainted" questionnaires (Table 2.3). Bring them with you this week.

TABLE 5.1 Sample physical geography feedback questionnaire

I would very much appreciate your taking a few minutes to answer these questions as fully and honestly as you can. Your answers will be taken seriously and will enable me to make changes in the future. They will also remain anonymous, so please do not put your name on them.

1. What in particular do you like about the labs? ***Note:*** If I am doing something particularly well, I would like to know about it so that I can continue doing so in the future. Please be as specific as you can.
2. What specifically do you dislike about the labs or the way I teach them? Again, I would like to know so that I can accommodate your ideas in future labs.

More specifically:

3. Lab quizzes
    a. Are they fair in content? If not, please specify.
    b. Are they graded fairly? Again, if not, please specify.
4. Do you feel that I manage to connect the labs well with the lectures and readings? If not, how do you think I might improve?
5. Do the movies and PowerPoint slides help in understanding the course material? Would the time be better spent doing something else?
6. Any other comments?

TABLE 5.2 Sample human geography feedback questionnaire

I would appreciate feedback about the recitation sections. I do have some flexibility in the way the recitation section is organized, and I want to address your needs and reactions. If you could take a few minutes to answer these questions, I would be grateful. I want honest, anonymous, unvarnished opinions, so please do not put your name on this.

1. What do you like about the recitation section so far?
2. What do you dislike about the recitation section so far?
3. Is the recitation section meeting your expectations of the course? Please elaborate.
4. What changes would you like me to make in the teaching and/or organization of the recitation section?
5. Is the feedback you receive on assignments/exams helpful? How could it be improved?
6. Do you feel comfortable participating? Please explain.
7. Do you feel that you are developing a sense of the discipline of geography? If your answer is no, do you have any suggestions as to what I might do in order to help you?

TABLE 5.3 Feedback questionnaire question bank

Organization

- Are there organizational or style changes which you would like to see in this section? Let me know.
- Is the lab taught at a level that is difficult/fast-paced, just right, or too easy?
- Do you feel that the topics in tutorial are covered too quickly, too slowly, or at the right pace for your understanding?
- Are you developing a sense of what a physical geographer does? If not, what else can I do to help you understand the field's advantages and limitations?
- *(For the daring!)* What else would you like to add that may help me teach more effectively? Remember you are paying a lot of money for this education of yours. You deserve to get the best education you can for the money you spend. State your case.

Examples, Explanations, Demonstrations, and Visual Aids

- Do you feel you have a good grasp of the concepts we have covered?
- Have you found specific examples cited during tutorial discussions to be helpful in understanding the subject being discussed?
- Does the lab help you with completing the assignment?
- Do the lab demonstrations help you in completing the lab assignments?
- Do you feel that I give too little, just enough, or too much guidance/instruction at the beginning of the lab sessions? Please explain.
- Are the handouts useful? Would you like more?
- Is my use of PowerPoint helpful, or would you rather I use the board or some other media?

(continued)

TABLE 5.3 Continued

Classroom Interaction

- What do you think of the class atmosphere? Is it too formal or informal; standard or unusual; am I overbearing or easy to get along with?
- I know it is not always easy to ask or answer questions out loud, but do you feel relatively comfortable speaking up in class? If not, what would make it easier?
- Do you feel comfortable asking questions in lab?
- Does working in groups (or as a class) and discussing the material during the lab help you understand the material better? Also, does this help you understand questions on the lab better?

Grading and Feedback

- Are the labs graded fairly, keeping in mind their weight in the overall grade for the course?
- How can I best give you feedback on your lab work? Which is more helpful: the written feedback on the lab itself or the oral feedback the following week? Please explain.

Instructor Qualities

- Do you feel comfortable coming to my office?
- Do you have conflicts with my office hours?
- Should I have additional office hours?
- Have you had trouble getting in touch with me or making it to my office hours? Please specify days/times?
- What do I do that is plain annoying or distracting? What do I do that you like and is different from other instructors you may have had?
- What do you like about the lab session or the way I presented it?

Note: Created from questions used by geography instructors.

Questions to Think About in Advance

- Having reviewed *students'* objectives and aspirations for the class (from the "getting acquainted" questionnaires), what do you think are the key issues on which *students* would like to give you feedback?
- Having reviewed *your* objectives and aspirations for the class, what are the key issues on which *you* would like to receive feedback from your students?

Which questions from the sample feedback questionnaires and question bank (Tables 5.1–5.3) most clearly address your issues and those of your students? (Select four to seven questions.)

- What question(s), if any, might you need to add to fill in any concerns not already covered?

TYPES OF FEEDBACK

There are many ways to obtain feedback from students. A chance encounter at the grocery store may elicit a surprisingly large amount of information. However, do not be tempted only to take the opinions of the most vocal students to represent those of the class as a whole. When you are trying to poll everyone, here are some things you might try:

- *Minute papers*—stop a few minutes before the end of class and ask the students to anonymously write on a spare piece of paper the answers to one or two probing questions (for example "Can you summarize the main idea from today's class in one sentence? Is there a concept from today's class that you feel needs further clarification?" etc.) and then have them turn in their responses to you.
- *In-class observation and feedback*—ask a peer to sit in on your class and talk to you about it afterward. You can then return the favor. This is a great way to pick up teaching tips. If you think there are things the students are not telling you, you might even try leaving early and letting your colleague talk to the students about the lab/recitation meetings and any changes they feel would be useful. In either case, areas to discuss could include your:
  - organization
  - use of examples, explanations, demonstrations, and visual aids
  - classroom interaction and rapport
  - grading and giving of feedback
  - personal teaching qualities
- *A feedback questionnaire*—this is often a very good place to start, which is why it is the focus of Activities 5.1 and 5.2.

NOTES ON CONSTRUCTING A FEEDBACK QUESTIONNAIRE

It may help to consider the following issues when constructing a feedback questionnaire.

- *Objectives of the questionnaire*—start by considering the objectives and format that useful feedback questionnaires might take. This will enable you to construct a questionnaire appropriate to your class situation.
- *Subject of questions*—your questionnaire should take into account the original desires of the students, and questions should be designed to establish whether original student aspirations (from your "getting acquainted" questionnaire developed in Chapter 2) are fulfilled. You need to emphasize that the questionnaire is designed to provide helpful feedback to you and is not in any way connected with other aspects of the course, for example the professor's performance during

lectures if you are serving as a teaching assistant (TA) in a large class. In this respect, the focus should also be on aspects of the course over which you have control (such as the lab or recitation environment and presentation of material), rather than the material presented in the main lectures. You should also focus on the issues that are most important for you (for example "Do you feel comfortable participating in the class discussions? Are my explanations clear, etc.?"). Do not try to cover everything.

- *Number and clarity of questions*—be sensitive to the amount of time students will have to respond when deciding how many questions to ask. Think about the logical order for questions and their clarity. Have someone else read them through to check for ambiguity.
- *Open-ended versus forced-choice answers*—in terms of question format, the general consensus from your peers is that open-ended questions, followed by a request for clarification ("If my explanations are not clear how might I improve them?"), provide the most useful information, in comparison to multiple-choice-type formats, which tend to constrain student answers.
- *Consistency among TAs*—if you are serving as a TA, you might want to discuss your personal teaching objectives with other TAs assigned to the same or similar courses and construct one questionnaire that can be used in all sections. While reviewing objectives and constructing the questionnaire, TAs teaching parallel labs/recitations of the same class often find it helpful to work together. Even TAs who are alone in a course find it useful to work in groups to pool ideas.

REFLECTIONS ON GETTING STUDENT FEEDBACK

Obtaining student feedback is an ongoing and essential process in all good teaching. It is the only way to accurately judge the progress of the class. Student feedback should be built into all aspects of the course. While it can be a little daunting to ask for your students' feedback on your teaching, the overwhelming consensus of teachers is that it is actually a very positive experience. Consider the reflections of Jay, a TA for Introductory Human Geography:

> There seems to be a general consensus that I am doing a pretty good job of explaining and expanding on the ideas and concepts which (the professor) mentions in lecture. Many students feel that lecture material is covered so quickly that they often do not always get "the point," so I am glad many think I am helping them understand what's being discussed. Several students commented that they feel comfortable participating in class discussions because I never make them feel silly if they do not say

> "the right thing." Most feel that I do not rush through the material, but that I cover it at an appropriate pace for their understanding. A number of students like the way the class has a relaxed feel to it, that is they do not feel uptight or threatened. I even had one student write, "the teaching is probably the best I've had from a TA so far."

In case you still have any doubts, here is what Trudy, a TA in a third year geography class, said after it was all over:

> One student wrote at the top of the form that she really appreciated that I asked for their input early on, and that she wished some tenured professors would do the same. That made the risk of criticism worthwhile.

Activity 5.1: Creating a Feedback Questionnaire

Use the information covered in the section above, "Notes on Constructing a Feedback Questionnaire," as well as Tables 5.1–5.3 to create your own feedback questionnaire. If you like the questionnaires in Tables 5.1 or 5.2, use those. They have been tried and tested, and there is no sense in reinventing the wheel if these already work for you. However, think about any questions you might wish to add or delete.

Activity 5.2: Administering Your Questionnaire in Lab or Recitation

After you have created your questionnaire, administer it in your class this week. Allow plenty of time for its completion. Emphasize the questionnaire's importance to you, its purpose and its anonymity. *Remember!*—if you are a TA, request permission from your course professor to do this. You may also wish to consider administering your questionnaire online if you have a lot of students. There are now many freely available online quiz packages for doing this.

You can share your experiences with other teachers and the facilitator at the next meeting.

CHAPTER 6

Responding to Student Feedback

Teresa Dawson

OK, take a deep breath and examine the feedback your students have given you. Is it what you expected? Did you realize you say "um" twenty-five times in a sentence? Have they given you some helpful suggestions? The objective of this meeting is to share both what we do well, and, if we wish to, what we could improve on. Do not be intimidated by this whole process. If your students understand that a feedback exercise is conducted because you are genuinely interested in improving what you do, they will usually be fairly gentle in their approach. It is not likely that their comments will be negative; rather they will make helpful suggestions. Overall you should be pleasantly surprised.

Things to Do Before the Meeting

- List for yourself the positive changes you would like to make to improve your teaching effectiveness.
- Find one additional/innovative idea from an article in the *Journal of Geography in Higher Education*, the *Journal of Geography*, or other geography journal that you could try in your class. Bring it to the meeting with you.

Questions to Think About in Advance

- How will you respond to your students' feedback this week?
- How could you adapt your journal idea to your particular classroom situation?

THE IMPORTANCE OF FEEDBACK

It is extremely important for your students' morale, as well as your own, to respond and let them know you are responding to their feedback as soon as possible. That is why Activity 6.1 asks you to think about how you will deal with your students' feedback from last week. But in order to be able to act, you first have to analyze and interpret the data. How should you respond? What are the appropriate changes to make? What should you do if you have a bimodal distribution of responses?

NOTES ON RESPONDING TO STUDENT FEEDBACK RESULTS

When responding to student feedback, think about the following issues:

- *Getting some distance from student comments*—it is best to wait at least until you get back to your office before you look at student comments. Do not read them in front of your students in case you get emotional. You might also want to wait until you have collected data from all your sections so that you get a better overall impression. When you do look at your data, try to do it in a relaxed fashion—when you have time to study them carefully and you are in a relatively stress-free environment.
- *Always respond the very next class*—students at universities and colleges are frequently asked to give feedback to their teachers. What they are less used to is a teacher who actually responds to what they have said. Therefore, if you really want students to give you feedback on future occasions, it is vital that you respond to them immediately. Others in the class, who did not answer your questions, or did not do so seriously, will be more likely to join in next time around.
- *What is appropriate to change?*—just because a student has suggested something, it does not mean you have to do it. Weigh the options carefully. As discussed in the last meeting, try to avoid asking questions about issues over which you have no control or that are not up for negotiation. If such issues come up, however, it is important to explain again why you cannot, or feel it is best not to, change current practice.
- *When do you need more data?*—fifty percent of your students think you go too quickly, fifty percent think you go too slowly. Fifty percent like the fact that you tell stories about your childhood; the rest would prefer you to write the equations for their assignment on the board. Bimodal distributions of student responses are some of the hardest to interpret and this is a case when you need to obtain more data. For example, what are the background characteristics of the students who think you go too slowly? Is your class split between majors and nonmajors? How can you accommodate their different

needs? Can you supplement the slower students' learning in your office hours? Is there any correlation between the pace the class is requesting and overall student performance on a test/assignment, or could everyone do with a little more explanation? Similarly, why do some students like the stories about your lost youth? Do they feel the stories distract you from giving them too much material to be responsible for on the test, or do they feel you use the anecdotes to help them relate to and better understand a difficult concept? When in doubt, first ask yourself what you know about the different characteristics of your audience, and then ask the students for more information.

- ***Getting another opinion in the department***—if you still feel you would like extra reflection on your feedback, and feel uncomfortable "sharing" in front of a large group, peers are an excellent source of nonthreatening input. Alternatively, make an individual appointment with the instructor of this course or a member of the faculty whose teaching you respect. Whom you choose is entirely up to you. You may wish to see your adviser, the professor you work for, someone else you take a class with, etc.
- ***Your university or college teaching and learning center***—Seek advice from your local center if at any time you are having problems:
  - interpreting the feedback you are getting from your students
  - designing a feedback questionnaire for your class
  - discussing feedback issues within the department

  Staff there will respond to your individual needs on a confidential basis, and they are often highly experienced at, and enjoy, working with graduate students and early career faculty.

THE ISSUE OF STUDENT MOTIVATION

When you review the feedback from students in your class, it is likely you will discover that some of your most important concerns are also theirs. One such joint concern is the issue of student motivation. While there is no magic formula that enables you to motivate your students—motivation comes from within—there are certainly ways in which you can facilitate them in motivating themselves. How is this best achieved? In the spirit of the feedback exercise we have just completed, we might start by asking the students themselves. Students have found the following activities to be among the most helpful in developing self-motivation.

- ***Acquiring information about the subject from a variety of sources***—the more students know about a subject, the easier it is for them to develop an interest in it. Try finding out which sources of information are most appealing to different individuals in your class, and see if each can contribute something to the subject at hand. For example, students could learn about the geography of slavery through written narratives, personal research projects, a review article in the *New York Times*, role playing, and field trips to appropriate sites, as well as through the more conventional means of textbooks and your lecture notes.

- ***Tying new information to old bodies of knowledge***—students may find that geography concepts and theories take on a new interest when they are seen in relation to present issues. Consider what is currently being reported in major newspapers that directly relates to your topic for the week and link it in.
- ***Making information personal***—relate the topic for the week to matters that are of real concern to the students. For example, if you are talking about deindustrialization and jobs moving abroad, find out if any of the students' families have experienced unemployment as a result of plant closings in the local area. Have the individual (or the class) research the story behind the particular example.
- ***Using new knowledge in an active way***—ask students to raise questions of their own about the readings or about what you say in class. They could hand the questions in at the end of class, and you could pick a few to respond to or to have the class discuss next time. Suggest students look for newspaper/periodical articles that illustrate what you have been discussing for the week. If a challenging question is raised, ask a group of students to research the answer.
- ***Being self-aware with regard to learning***—encourage your students to be self-aware when it comes to their learning. How do they study/learn best? How can they adapt most easily to, and get the best out of, your teaching style?
- ***Having definite tasks***—decide what needs to be done, and when, in order to achieve your educational goals. Convey your rationale to the students and then, within reason, allow them to participate in some of the course policy decision-making. If a task is too daunting for an introductory class, help the students break it down. Model the geographer's problem-solving process for them and clearly indicate the rationale behind each part. As the semester goes on, students will be able to take more initiative in breaking down problems for themselves.
- ***Undertaking small group work***—working in small groups is an excellent way for students to take ownership of the material, to teach it to each other in a meaningful way, and to bring a subject to life for those members who might be less interested in a particular topic. If you feel such collaborative student work is useful, then it is important to model it in class. Try assigning teams for the semester, charging each with particular responsibilities, such as reporting back on the readings, solving a particular problem, or undertaking a group project. It is likely that the desire not to let down peers in the group will

significantly increase participation of all members in the assigned activity. Once students begin to know something about a subject, their interest level will rise. Also, once they see the benefit of group collaboration, they are more likely to extend this behavior more broadly to study groups outside the class.

- ***Individual tutoring and use of office hours***—encourage students to tutor each other on topics they find difficult. Also, get students to talk to you as much as possible. There are often powerful forces of inertia that prevent students from coming to office hours (walking all the way across campus to your building, climbing up to the third floor, etc.). Yet, those who do come often find themselves more motivated and more successful in completing a particular class activity. Help students to overcome this inertia by building in an office visit for every student early in the semester. Once students have been to your office once and found the visit useful, they are more likely to come again.

Activity 6.1: Reflections on Student Feedback

Please hand in the following to the meeting facilitator:

1. ***A copy of the feedback questionnaire you used.***
2. ***A one-page commentary on the feedback you have received.*** Write a paragraph each on (1) your students' aspirations for the course taken from the original "getting acquainted" information you solicited at the beginning of the semester (see Chapter 2), (2) your strengths according to your students, (3) any weaknesses your students identified (you do not have to tell the facilitator all or any of the grim details if you do not want to), and (4) what you might do differently as a result of the feedback. For those unassigned, please hand in (1) a copy of the questions you would like to ask, (2) a written statement of how you think the students might respond to them, and (3) your reflections on these possible responses.

CHAPTER 7

Access for All
Issues of Classroom Equity

Teresa Dawson

Do you feel you are walking a tightrope when it comes to issues of equity in your classroom? Do the words "political correctness" throw you into confusion? Should you call on the only African-American student in your class to represent "the African-American perspective" on environmental racism? Should you expect your female students to understand feminist geography? How can you eliminate homophobia and ageism in your classroom? What terms should you use to be respectful to native peoples (and how does the answer to this question vary from state to state in the U.S. and from province to province across Canada, as well as evolve over time)? How should you prepare for the needs of students with disabilities who might wish to take your class and still be fair to other students? What problems might your international students be having adapting to your classroom context?

Teachers face such questions every day. They are often concerned that they might offend students in their classes, and are anxious to avoid any such offense, but they may be unsure as to how best to deal with some of the more complex classroom diversity issues. As with many issues, often the most useful approach is to ask the students concerned how they feel. Accordingly, the objective of this week's meeting is to discuss a case study written from a student's perspective, and to use this case study to explore the broader issues of diversity in universities and colleges.

Things to Do Before the Meeting

- Read the following Josephine James and John Close case studies written for this meeting by Nancy Evans and Jennifer Johnson (Figures 7.1 and 7.2).
- Carefully consider and answer the questions about these case studies found below.

Questions to Think About in Advance

- The Josephine James case study raises some important issues about our attempt to accommodate the needs of students with disabilities in our classes. How are these issues similar to or different from the issues that arise when we consider the needs of other frequently excluded groups such as the African-American students in the John Close study?
- How specifically might you go about creating a classroom climate that accommodates the needs of *all* the students involved?
- What additional questions/issues do you have concerning diversity in the university or college classroom that you might want to discuss?

CASE STUDIES AS A WINDOW ON DIVERSITY ISSUES IN THE CLASSROOM

Getting to meet a wide variety of people is an important and extremely rewarding part of what it means to get an education at a university or college. Therefore diversity should be an integral part of what we do in the classroom, providing opportunities for the enrichment of learning of both teacher and student. While this may generally be the case at your institution, sometimes there are issues that arise out of contact with people who are different from ourselves, or from the majority of our students, that we may feel ill-equipped to deal with and might like guidance on. Yet we may feel awkward about voicing our concerns or asking genuine questions for fear they might be misinterpreted.

Experience at other institutions suggests that one effective way of allowing people to openly discuss their questions concerning diversity issues, as well as better informing them about such issues, is to use a case study approach. This approach begins by presenting the narrated experience of an individual who feels marginalized in a particular classroom situation (it may be an account of an actual experience or a scenario based on the experiences of several individuals). The individual who feels in the minority may be the teacher or a student. The case then proceeds by asking the group, "Given this scenario, what can be done and how might the situation best be handled?" Finally from the discussion of the particular case, broader questions then arise for consideration by the group.

BEING PROACTIVE ABOUT DEVELOPING YOUR OWN TEACHING RESOURCES DIRECTORY

Your department should provide you with a list of useful numbers in a general directory pertaining not just to diversity issues, but also to teaching in general. Or, you can start to collect such a list as you go along. For example, have the number of your institution's crisis center handy; it is extremely important should you be faced with a student in crisis (that is not the time to have to look up the number).

Activity 7.1: Case Studies of Classroom Diversity

The following are case studies written for geography teachers by Nancy Evans and Jennifer Johnson. Keep their questions in mind as you read the cases. Try putting yourself in the position of both the teachers and the students involved. Then ask yourself what you would do if either of these situations arose in your class.

In the case of Josephine James (Figure 7.1):

1. What could Josie have done to better handle this situation?
2. Should Josie have anticipated that she might have a disabled student in her class and taken this into account in planning her syllabus?
3. Should Josie have adjusted her syllabus to accommodate Lisa? Would such an adjustment have been fair to Lisa and to the other students in the class?
4. Should Josie have handled the class differently? For instance, should she have asked the students engaged in the side conversation to stop? Should she have asked Lisa to express her concerns in class?
5. How might Josie have handled her postclass discussion with Lisa more effectively?
6. Is dropping the class the best solution for Lisa?

In the case of John Close (Figure 7.2):

1. What are the pros and cons of John assigning all of the African Americans to the panel on the underclass?
2. Do you think that Gwen was justified in her angry reaction to the group assignment?
3. What should John tell Steve as the only white member of the group?
4. Should John discuss with anyone Steve's concern that the other students might be biased? If so, with whom? Would it help to talk to Steve alone, to all four group members together, and/or to the entire class?

FIGURE 7.1 Profile of Josephine James by Nancy Evans, Department of Counseling and Rehabilitation Education, Penn State University

Josephine (Josie) James is a new teaching assistant (TA) who has been assigned a lab section of physical geography. Josie is very excited about this opportunity since she has always wanted to be a university instructor, and has lots of ideas for making physical geography exciting and relevant for her students. Josie has spent the summer prior to beginning graduate school working on her class. She firmly believes that hands-on learning is the only way to make the concepts of physical geography real to beginning-level students. As a result, most of her class sessions involve active participation and work with maps. She has units in which students create their own maps that illustrate the distribution of natural resources, areas of environmental mismanagement, etc. Josie has also planned a number of field trips so students can observe firsthand the effects of environmental pollution, forestry management practices, and other issues related to the environment. Many of these field trips involve a significant amount of hiking in rather rough terrain, but Josie figures that undergraduates will enjoy the challenge and the change from sitting in a classroom every period, and that they will learn more by actually interacting with and experiencing the environment than they ever would just listening to lectures and reading about these issues.

Josie approaches her first class with a bit of apprehension, but mostly with excitement. She is eager to get to know her students and to share her excitement for physical geography with them. She gets to class early and greets the students as they enter the room. Close to the starting time of class, a young woman enters who obviously has trouble walking. She uses crutches and has braces on both legs. It takes her some time to enter the classroom and take a seat. Once she is seated, Josie also sees that the young woman has limited use of her hands. She has brought a tape recorder to record the class because she cannot hold a pencil to write. She asks Josie if this is acceptable to her and Josie says she has no objections.

Josie distributes a syllabus to the class that outlines the activities and assignments she has planned. As she begins to go over the syllabus, she notices that this young woman is getting more and more distressed. Josie realizes that many of the activities that she has planned will be impossible for this student. She also notices that other students in the class keep glancing at this young woman and that a few side comments are shared between students sitting in the back of the room. Josie is very uncomfortable with the entire situation but chooses to ignore it since she does not want to call attention to the young woman. She finishes the review of her

(continued)

FIGURE 7.1 Continued

syllabus as quickly as possible and asks in a perfunctory way if anyone has any questions. Since no one does, she gives the assignment for the next week and lets the class go early. She wants to talk to the young woman with a disability but does not know what to say.

The young woman, whose name she finds out is Lisa Maxwell, waits until all the other students have left and then approaches Josie. She asks Josie if she received a notice from disabled student services informing her that Lisa would be in her class and that Lisa has cerebral palsy. Josie tells Lisa that she did not get the note and perhaps it had gone to the professor in charge of the class. Josie apologizes to Lisa for the mix-up. Lisa tells Josie that she plans to drop the class since it is evident that she cannot do the required work. Josie is flustered but tells Lisa that might be best. Josie can see that Lisa is close to tears, and she again apologizes for the uncomfortable situation. Lisa makes her way out of the classroom. Josie feels terrible but does not know how else she could have handled the situation.

FIGURE 7.2 Profile of John Close by Jennifer Johnson, Department of Geography, Penn State University

As a new TA for the department's undergraduate social geography course, John Close was determined not to do the same old things in class. He remembered how bored he had been as a student and had tried to come up with a variety of approaches to the different issues the class needed to cover. Making his lectures interesting and getting students to participate whenever possible were a critical part of his course plan.

One thing John liked about his class was that it was relatively small and included a mix of students: three were African American and altogether roughly half of them were women. As far as John was concerned that was good because he wanted to encourage perspectives other than that of the traditional white male. He also wanted to include discussions on topics of current interest such as the urban underclass and the racist right. So when he came up with the idea of having student groups take turns being on panels, researching topics, and discussing them in front of the class, he thought he would really hit on something great. That would give him the opportunity to combine student participation and expose the class to multiple viewpoints.

After the first week of class, John decided to assign students to panels so that they could start working on the assignment. He divided the class into groups of four and handed out discussion topics, assigning one white student and the three African Americans to the group investigating the underclass.

The next day one of the students in the group working on the underclass stopped by during office hours. The woman, whose name was Gwen, felt that she and the other two African Americans had been singled out of the group; she also said that she resented the implication that "all blacks must know about the underclass" and they should "speak for their race." In addition, Gwen was really interested in pursuing another topic. She proposed that the groups be changed and she be allowed to work on another issue. John disagreed because he did not feel like he had "singled out" any of the students, and one goal of the project was to research and prepare an argument on a topic that might not necessarily be of interest to the individual concerned. His decision was to keep the groups as they stood. Gwen left still fuming.

Later that week another person came to see John. This time, however, it was Steve, the only white student in the group researching the underclass. Steve also had concerns about being in the group, but they revolved around feelings of discomfort and alienation. After a few study sessions he was feeling left out because the other three students talked all the time and he did not feel comfortable giving his opinion on any of the issues. One reason was that the other students spent a lot of time discussing racism and the types of discrimination that occur in current society. Another reason was that he thought they might be presenting a biased view of the issue because they were minorities, thereby potentially giving the rest of the class a distorted picture. Steve was not sure where the project was going, was unenthusiastic about the topic, and did not think his staying in the group would help anyone. After asking to be switched to a different group Steve left, leaving John confused and worried that what had started as a simple teaching technique had become a problem for everyone.

CHAPTER 8

At Home in the Classroom

Developing Your Own Teaching Philosophy and Portfolio

Teresa Dawson

What kind of a teacher are you? What is your style? This week's meeting is designed to refine what you have learned over the semester, to talk a little more about the theory behind teaching, to help you gain more control over the direction of your classes, and to explain why some things work and others do not (in case you were wondering). Its objective is to facilitate your moving from "effectiveness" to "belonging" in the classroom—or in geographical terms, to create a sense of place out of your teaching space.

One important way of deciding who you are as a teacher is to identify your teaching philosophy. Indeed those of us who are on the job market are very aware of the need to be able to present and defend a coherent teaching philosophy. While your philosophy may change and develop during graduate school, it is a good idea to have already begun to formulate one, if for no other reason than you can then reflect upon and refine it. In addition, it is useful to be able to summarize your philosophy in a form suitable for a job interview, resume, or more broadly for your teaching portfolio. Your statement of teaching philosophy can also help you create a development plan for your teaching so you can consider how you would like to improve your abilities and what concrete steps you can take toward reaching these goals.

Although it may seem too soon to consider writing a statement of teaching philosophy or assembling a teaching portfolio, there are three reasons for getting an early start with Activities 8.1 and 8.2. First, reflection is a key to learning, and taking stock of our strengths and weaknesses helps us clarify where to invest our time and effort wisely. Second, assembling a teaching portfolio involves collecting materials and data of many sorts, so it is good to start gathering this evidence sooner rather than to realize too late that you did not keep something you needed. Third, the principles for creating a good teaching portfolio are also of value in creating an effective dossier for tenure and promotion. That is, the questions in Activity 8.2 about articulating your greatest strengths as a teacher can apply just as well to research, service, and all your other scholarly and academic work.

Things to Do Before the Meeting

- Write a short statement of your teaching philosophy, also called a "teaching statement" or an "approach to teaching." Modify this into a form that you do not mind others seeing and bring it to class *unnamed*. *Do not* share it with anyone. Your statement will form the basis for this week's meeting and for Activity 8.1.

Questions to Think About in Advance

- What does good teaching mean to you?
- What are your greatest strengths or accomplishments as a teacher?
- What concrete examples and evidence do you have of these strengths and accomplishments?
- How would you improve your teaching in the future? Can you be specific about the sorts of strategies, ideas, practice, readings, or activities you would pursue to make improvements?
- What are your career goals and how does teaching fit into them?
- Might a teaching portfolio contribute to the achievement of those goals?

THINKING ABOUT YOUR TEACHING PHILOSOPHY

Now that most of you have had several weeks of classroom experience and feedback, we can reflect on how you are adapting the generic, theoretical methods for effective teaching from Chapters 1 and 2 to form practical, discipline-specific approaches to teaching in geography. You should now be in a better position to look at the methods and principles more critically and reflectively. For example, what works in geography teaching and what does not? Are there any significant differences between human and physical geography teaching strategies? What does it mean to you to be a good teacher? As we have discussed before, there are many good and successful teachers and every one of them has a different style that fits their personality. With this in mind, we should be ready to play the "Teaching Philosophy Game" (whereby we can guess from what is said which teaching philosophy statement belongs to whom). Essentially, in this game each participant picks a random unnamed philosophy out of a hat and reads it to the group and then others guess to whom it belongs. This demonstrates the point that philosophies are personal and that they are successful when they reflect the person (in other words, you should be able to guess who wrote which one).

YOUR FUTURE PROFESSIONAL DEVELOPMENT

Whatever your career goals, chances are you will end up using many of the skills you have developed in this series. Whether in academia or in a volunteer or corporate situation, you will always have to explain concepts clearly to others and make an argument for your point of view. The better a teacher you are, the better you will be able to accomplish these tasks. If you are going on to a career in academia, it is very likely that you will have a position that is made up of some component of research, teaching, and service, and your pay and your advancement will therefore depend in part on your ability to document your teaching effectively. Increasingly, the teaching portfolio is the document of choice for institutions that wish to assess faculty fairly and equitably. For this reason, your teaching portfolio (and the work you have done in this class) may help you get a job, but it will not stop there. The teaching portfolio is a living document that will grow and change with you throughout your career. Learning how to construct one now is a great investment in your future.

Activity 8.1: Writing Your Teaching Philosophy

Your statement of teaching philosophy is the first and most important document in your teaching portfolio. Please hand in a copy of your teaching philosophy narrative and a list of evidence and documents you would use to support it (for example, if you were sending it for an award or a job application). As you write your narrative, think carefully about practical examples of how you are able to implement your philosophy in specific classroom contexts and use those examples to illustrate and document the claims you are making.

Using the term "philosophy" sometimes confuses people into thinking they should write a statement exploring the epistemological foundations of learning based on the findings of educational psychologists over the last century. Instead, it is perhaps more useful to think of writing a statement like "My strengths, accomplishments, and background as a teacher," focusing on:

- What are your most important strengths, characteristics, or accomplishments as a teacher?
- What is the nature and scope of your experience and training as a teacher?
- What are your key teaching goals generally or in the context of particular classes, topics, or audiences?
- How will you develop your abilities further in the future?

Activity 8.2: Creating a Teaching Portfolio

A teaching portfolio, analogous to an art portfolio, is a well-established device used to improve and assess teaching across North America and beyond. The idea is that such a portfolio is better able to demonstrate the depth and breadth of your teaching skills than is a computer-scored evaluation form, for example. This activity is designed to help you through the process of creating your own teaching portfolio.

The portfolio essentially consists of two main parts: (1) your narrative teaching philosophy and (2) documentation that supports that philosophy. Much of what we have done in Chapters 1–8—particularly your analysis of your mid-semester student feedback—will contribute to your portfolio. Make a note to yourself to continue collecting documentation, and begin now by writing or revising your teaching philosophy using the matrix in the workbook to help organize your thoughts.

What Is a Teaching Portfolio?

As Seldin (1991, 3) says of a teaching portfolio, "It describes documents and materials which collectively suggest the scope and quality of a teacher's performance." At many universities the processes of annual reviews, tenure, and promotion are beginning to place greater stress on faculty submitting teaching portfolios—personalized collections of materials that document teaching effectiveness—along with their *curriculum vita* and annual reports. While many faculty are already aware of the need for such a portfolio, some are still confused as to what exactly a portfolio is and what it should contain.

To begin, write an answer to the following question: What do you want your teaching portfolio to do for you? Will you use it for job hunting, for annual review, for a tenure and promotion review, or for other reasons?

Where Do I Start?

Once you have answered the previous question, consider two more:

1. What major claims do you make about your teaching? (What sets you apart as a teacher? What do you think are your most important characteristics as a teacher? What are your key teaching goals?) Try to limit yourself to three.
2. What types of instructional methods, materials, and techniques do you use to support your teaching goals? Include any that are particularly innovative.

It may help to focus on either (a) the course you have enjoyed teaching the most or (b) the course you have taught most often.

What Possible Data Sources Should I Consider?

The real secret of assembling a successful portfolio is knowing whom to ask for what. Seldin (1993) lists the kinds of materials that one might include in a teaching portfolio. The specific items you select will depend on your particular teaching assignment and activities. It will also depend on your potential to generate opportunities for collecting materials suitable for inclusion in your portfolio. Such opportunities might include discussing how you can combine your teaching and research interests. Some ideas for sources of data are given below. As you read through them, check those materials that you would be able to include in your teaching portfolio and star those you had not previously thought of.

Data from oneself. Self-analysis and self-reflection are far too often overlooked in the assessment of teaching and learning, yet they are central not only to the processes of assessing teaching but also to improving it. Thus they are an essential part of your teaching portfolio.

Faculty members can provide their own perspective on virtually every aspect of instruction. Self-reports should be primarily descriptive as opposed to evaluative—what were you trying to do, why, how, and what was the result? Consequently, these self-generated documents will more easily reflect development than those from other sources. Self-reports should also be compared with data from other sources. Because feedback that provides new information is most likely to produce change, it is by virtue of such comparisons that personal growth and improvement occur. Data from oneself might include the following:

- a list of courses taught, with brief descriptions of course content, teaching responsibilities, and student information;
- a statement of one's philosophy of teaching and factors that have influenced that philosophy;
- examples of course materials prepared and any subsequent modifications that were made to accommodate unanticipated student needs;

- a sample syllabus or lesson plan;
- a record of teaching discoveries and subsequent changes made to courses regularly taught;
- a description of efforts to improve teaching (*e.g.*, participating in seminars and workshops, reading journals on teaching, reviewing new teaching materials for possible application, pursuing a line of research that contributes directly to teaching, using instructional development services, and contributing to a professional journal of teaching in your discipline);
- evidence of reputation as a skilled teacher, such as awards, invitations to speak, and interviews.

Data from others. Obviously, different people can provide different kinds of information about your teaching. For example, it is probably counterproductive and inappropriate to ask students about the breadth and completeness of an instructor's content knowledge since, from their point of view, such expertise should be assumed. The more obvious and appropriate judges of this information would be department colleagues. Likewise, such colleagues are usually not good judges of whether an individual is prepared for class, arrives on time, or is available for office hours. Clearly, getting the right kinds of input from each group of individuals is what will give your portfolio its strength and depth.

As the immediate beneficiaries of teaching, students are in an ideal position to report and comment on a number of factors, such as which instructional strategies helped them learn the most, whether the instructor came prepared to class, was available during office hours, or provided useful comments on papers. Other data that only students can report involve any change in their level of interest as a result of taking the course, how much the course challenged them, and whether they felt comfortable asking questions. The most common ways of obtaining student feedback about these aspects of teaching include the following:

- interviews with students after they have completed the course;
- informal (and perhaps unsolicited) feedback, such as letters or notes from students;
- systematic summaries of student course evaluations—both open-ended and restricted choice ratings;
- honors received from students, such as a teaching award.

Other materials often referred to in the literature on teaching portfolios are the "products of good teaching." In a sense, these are a subspecies of the broader "data from students" category and might include the following:

- examples of the instructor's comments on student papers, tests, and assignments;

- pre- and post-course examples of students' work, such as writing samples, laboratory workbooks or logs, creative work, and projects or field work reports;
- testimonials of the effect of the course on future studies, career choice, employment, or subsequent enjoyment of the subject.

Colleagues within one's own department are best suited to make judgments about course content and objectives, the instructor's collegiality, and student preparedness for subsequent courses. Departmental colleagues can provide analyses and testimonials that serve as a measure of:

- mastery of course content;
- ability to convey course content and objectives;
- suitability of specific teaching methods and assessment procedures for achieving course objectives;
- commitment to teaching as evidenced by expressed concern for student learning;
- commitment to and support of departmental and instructional efforts;
- ability to work with others on instructional issues.

Data from colleagues could include the following:

- reports from classroom observations by other faculty or instructional specialists;
- statements from those who teach other sections of the same course or courses for which the instructor's course is a prerequisite;
- evidence of contributions to course development, improvement, and innovation;
- invitations to teach for others, including those outside the department;
- evidence of help given to other instructors on teaching, such as sharing course materials.

Remember, not all teaching takes place in the classroom. The most significant administrators who may not necessarily be included in the previous category are the department chairs. In their supervisory capacity, these administrators are generally well suited to make summary statements about overall performance over time. In doing so, they can help those who will read and interpret the portfolio by organizing and assimilating all the other information from various sources. It is also appropriate that these individuals draw attention to special recognition for teaching such as a university teaching award.

What Might I Include in My Portfolio?

What types of evidence do you currently have (or could you collect in the future) in order to demonstrate your major claims? Remember, not all teaching takes place in the classroom. Faculty do a great deal of out-of-class mentoring. When you consider the teaching you do, take the broadest definition of your contributions.

How Do I Select and Present the Material?

Selection. Clearly you should not put all the materials you have collected in a large box and send them as is to an unsuspecting department chair. Before you engage in the necessary process of selection, consider the following questions:

1. Why are you creating a teaching portfolio?
   - departmental teaching assignment decisions
   - merit assessments
   - job/grant application
   - self-analysis or reflection
2. Who is your audience?
3. What is the overall argument you wish to make?
4. What are the norms regarding length and depth of a teaching portfolio in your department/discipline?

Arrangement and presentation of components. A teaching portfolio is, and indeed should be, highly personal. There is no specifically recognized format at many universities. In addition, departmental guidelines may apply, so faculty should always consult their chair for the most recent information. In the most general sense, a portfolio is likely to contain a short reflective narrative followed by an appendix of supporting documentation. Beyond this, selection and arrangement should be done so as to best reflect the argument you wish to make. Take a few minutes now to begin planning your portfolio. Try using the organizational matrix found in Table 8.1 to help you.

Reflective narrative. This key piece of your portfolio includes the major claims you wish to state about your teaching and indicates how these claims support the case you are making. You will need to use specific examples that narrate your claims and give them flavor. For this you can draw on the "Data from oneself" section above.

Supporting materials, data, and documents. These elements are used to illustrate the claims and examples in your reflective narrative, and hence to support your overall argument. For this you can draw mostly, though not exclusively, on the "Data from others" section above. You might include, for example, a table of standardized course evaluations, as well as a sample lesson plan or syllabus. Supporting materials are most conveniently located in appendices. They need to be carefully selected so as not to be too lengthy (pick just the clearest example to support your point), and should be arranged and labeled for the convenience of the reader. Points made in the narrative should be referenced to specific pages or parts of the appendices if at all possible.

TABLE 8.1 Organizational matrix for aligning claims and materials in teaching portfolio

Major Teaching Claims	Specific Narrative Example of Claim	Supporting Data for Appendices
Source: Data from Oneself	Source: Data from Oneself	Source: Data from Others
EXAMPLE: *I encourage students to appreciate alternative viewpoints.*	*I always include an in-class debate over a controversial topic (e.g. XXX) in my syllabus. In these debates students are assigned a role and asked to argue from a perspective other than their own.*	*my class outline for the debate on XXX *student testimonials from my written evaluations

Checking your portfolio for balance. Once your matrix is complete, and before you write your final draft, check your portfolio for balance. In particular, make sure the "Data from others" comes from multiple sources (students as well as colleagues).

How Do I Get Feedback from Others?

Does your portfolio make the strongest case it can for your application, and does it reflect you as a person? As is the case for teaching in general, the best portfolios are those that are constantly revised and updated. Input from colleagues and friends can be invaluable in this process.

If you are working on this activity in a group, begin now by breaking into small groups. Each member of your group should take a turn and briefly explain the overall message they are trying to convey in their portfolio and the materials they intend to use to support that message. Other members should then make suggestions to strengthen the speaker's case. If you are working alone on this activity, discuss your ideas with peers, colleagues, advisors, or mentors.

What suggestions do colleagues have for strengthening your portfolio?

Putting Your Portfolio on the Web

Increasingly, faculty are making an electronic version of their teaching portfolio available on the web under password protection. While this is certainly optional, if you are comfortable with electronic media, or are in a field where new media applications are expected, you can probably appreciate the additional flexibility the electronic portfolio offers. Faculty are often excited by this new format, which enables them to present nonlinear ideas and to adapt their portfolio much more easily to different audiences.

Additional Resources

Anderson, E., ed. 1993. *Campus use of the teaching portfolio: Twenty-five profiles.* Washington, DC: American Association of Higher Education.

Describes the current form and use of teaching portfolios in twenty-five varied U.S. college contexts. Includes details of standardized forms and reflections on portfolio use from the institutions concerned.

Cashin, W. E. 1989. *Defining and evaluating college teaching.* IDEA Paper No. 21. Manhattan, KS: Kansas State University, Center for Faculty Evaluation and Development.

Cashin argues for a more comprehensive evaluation of teaching that considers a wider range of information from a variety of sources. Although he does not specifically discuss teaching portfolios, he establishes a useful guide for assembling one by defining seven aspects of teaching and discussing the most appropriate sources of information about each.

Edgerton, R., P. Hutchings, and K. Quinlan. 1991. *The teaching portfolio: Capturing the scholarship in teaching.* Washington, DC: American Association of Higher Education.

A good place to get started with background information. Gives detailed examples of how teachers demonstrate and reflect on changes in their teaching using examples of student work.

Shore, B. M., S. F. Foster, C. G. Knapper, G. G. Nadeau, N. Neill, and V. Sim. 1986. *The teaching dossier: A guide to its preparation and use*. http://www.caut.ca/uploads/teaching_dossier_en.pdf? (last accessed 15 September 2007). Ottawa: Canadian Association of University Teachers, 1986.

One of the key original reference pieces for those working in the area of teaching portfolio development. This work contains a list of forty-nine types of items that might be included in a portfolio. A good source of ideas for additional materials.

Shulman, L. S. 1998. A union of insufficiencies: Strategies for teacher assessment in a period of educational reform. *Educational Leadership* 46(3): 36–41.

Makes a good case for the use of portfolios and suggests they reflect both the teacher's efforts and the input of mentors or peers. Uses reference to the film *Stand and Deliver* (about a successful teacher) to illustrate his points.

References

Seldin, P. 1991. *The teaching portfolio: A practical guide to improved performance and promotion/tenure decisions*. Bolton, MA: Anker.

———. 1993. *Successful use of teaching portfolios*. Bolton, MA: Anker.

SECTION II

Promoting a Scholarship of Teaching and Learning in Geography

Michael Solem, Kenneth Foote, and Cary Komoto

The four chapters in this section were designed with two goals in mind. First, each chapter provides practical advice on pedagogical topics of enduring interest to geography educators. Second, the chapters illustrate some of the ways that geographers are researching the effectiveness of their teaching and sharing that knowledge with their colleagues. These practices exemplify the spirit of the scholarship of teaching and learning (SoTL) (Hutchings 2000).

Certainly, much of academic geography's renaissance in higher education and society can be attributed to the growth of geographic information systems (GIS), Global Positioning Systems, remote sensing, virtual globes, and other technologies for mapping and analyzing spatial data. In Chapter 9 "GIS and Mapping Technologies: Applications for Reasoning and Critical Thinking," Diana Stuart Sinton and Rich Schultz discuss the ways geographers can incorporate geospatial technologies to support analytical thinking in the undergraduate curriculum. Diana and Rich make clear that significant learning with GIS requires an understanding of geographic concepts and principles, a point often lost when the technology is viewed merely as a software application for displaying maps. The online activity for this chapter is designed to help professors who otherwise have little experience with technology make effective use of Google Earth and GIS for analyzing social and environmental phenomena.

Facing a class of students for the first time can be daunting for new instructors, and even veterans will admit to feeling butterflies stirring in their stomachs on the first day of class. At larger universities, the introductory course in geography can enroll hundreds of students packed into a lecture hall. The prospects of teaching in such a venue have led to many sleepless nights for new professors. For many instructors in this situation, the lecture is the fallback technique for teaching. Not only is it the most comfortable and familiar method, but lecturing also provides the instructor with a sense of control in an intimidating environment. But as Doug Gamble notes in Chapter 10 "Looking Beyond the Lecture: Promoting Significant Learning in Large Classes," it is possible to create lectures that feature a variety of engaging, in-class activities that make learning more active and reflective, while technologies such as the web provide the instructor with additional options for maximizing the effectiveness of large classes. Paired with this chapter is an online activity that can help you align course goals with active learning strategies in large classes.

Fieldwork is perhaps the hallmark of geographical teaching. In recent years the traditional field study has lost some cachet because of concerns over cost, liability, and time commitments. However, as Jennifer Speights-Binet and Doug Gamble show in Chapter 11 "Teaching in the Field," there are many opportunities to engage students in fieldwork appropriate for physical and human geography courses. The college campus itself provides a readily accessible venue for developing field exercises, and virtual technologies extend the traditional boundaries of field instruction to include places not normally within reach of many students. Jennifer and Doug make clear the many logistical issues involved in planning field studies, and the need to make these learning experiences accessible for all students. Mindful of the needs of students learning in the field, the chapter activity can help you design field exercises that will help your students achieve a variety of learning objectives.

In an interdependent world, there is a compelling need for undergraduate students to acquire international perspectives on issues such as economic development, environmental pollution, political change, and population growth. In Chapter 12 "Geography and Global Learning," Michael Solem argues that geographers can enhance the content of their courses by "internationalizing" teaching and learning. He outlines a variety of strategies for building intercultural perspectives into geography courses, pointing out the geographical dimensions of learning to become a global citizen. Michael's chapter highlights resources developed for the AAG Center for Global

Geography Education (CGGE), a project that links geography classes worldwide through a collection of online course modules exploring different geographical issues. The chapter activity can get you started with the CGGE modules and help you build teaching collaborations with geographers in other countries.

One of the recurrent themes linking all of the chapters is that teaching and learning issues are worthy of scholarly investigation. Consider the perspective of Randy Bass, one of the founders of the International Society for the Scholarship of Teaching and Learning:

> One telling measure of how differently teaching is regarded from traditional scholarship or research within the academy is what a difference it makes to have a "problem" in one versus the other. In scholarship and research, having a "problem" is at the heart of the investigative process; it is the compound of the generative questions around which all creative and productive activity revolves. But in one's teaching, a "problem" is something you don't want to have, and if you have one, you probably want to fix it. Asking a colleague about a *problem* in his or her research is an invitation; asking about a problem in one's teaching would probably seem like an accusation. Changing the status of the *problem* in teaching from terminal remediation to ongoing investigation is precisely what the movement for a scholarship of teaching is all about. How might we make the problematization of teaching a matter of regular communal discourse? How might we think of teaching practice, and the evidence of student learning, as problems to be investigated, analyzed, represented, and debated? (1999)

Questions about student learning can take many forms. One type is the "what works" question that usually takes the form of evaluating the effectiveness of a particular teaching approach or method. A second common type is the "what is" question about the learning process—how learning occurs, as opposed to considering only the outputs or end results of learning as evidenced by exam scores. There are many other forms of questioning in education (Bransford, Brown, and Cocking 2000), and just as in other forms of research the investigator must consider relationships between the question and the forms of evidence or data needed to arrive at answers. In any discipline, evidence of student learning questions might include results from exams, student essays, or projects, while techniques such as participant observation and interviewing can provide insight into the process of learning at formative stages.

The end result of educational inquiry is often a product that can be used by others. Traditionally in research, this is a peer-reviewed article, while in SoTL the product could be a peer-reviewed article, but it does not necessarily have to be. Other products could be a published essay by the researcher, a workshop for training faculty in new methodologies, a case study, or teaching materials that are demonstrably effective for improving student learning. One outcome of this dissemination is contributing to a body of knowledge about "what works" in improving undergraduate education in a discipline. In this sense, SoTL can not only improve the quality of your own courses but also promote systematic, peer-reviewed approaches to teaching and learning in forms usable by others. As you read the following chapters, please take note of the "SoTL sidebars" that appear in each. The SoTL sidebars connect each topic with some aspect of SoTL, such as publishing the result of an educational project in a peer-reviewed journal or developing an interview protocol to explore student responses to a problem-based learning activity. We hope the SoTL sidebars inspire your own thinking about how you can begin to study the process and outcomes of your teaching, and that you will use this book's online community-based web site as one outlet for sharing that information with others.

References

Bass, R. 1999, "The Scholarship of Teaching: What's the Problem?" *Inventio*, http://www.doit.gmu.edu/Archives/feb98/randybass.htm (last accessed 2 August 2007).

Bransford, J. D., Ann L. Brown, and Rodney R. Cocking, eds. 2000. *How people learn*. Washington, DC: National Research Council.

Hutchings, P., ed. 2000. *Opening lines: Approaches to the scholarship of teaching and learning*, Menlo Park, CA: The Carnegie Foundation for the Advancement of Teaching.

CHAPTER 9

GIS and Mapping Technologies

Applications for Reasoning and Critical Thinking

Diana Stuart Sinton and Richard B. Schultz

Geographic information systems (GIS) and mapping tools can be very effective in helping students develop critical-thinking and problem-solving skills. These technologies can also help students visualize data, evaluate hard-to-discern patterns, improve quantitative reasoning skills, and reach deeper levels of learning, especially that which is related to spatial thinking and reasoning (National Research Council 2005; Sinton and Lund 2007). However, students are unlikely to arrive at college with any understanding of spatial data, much less have any exposure to geographic concepts of scale, projections, or cartographic design. Thus, it is important to discuss the basics of data collection, organization, and analysis, together with map preparation, in all levels and types of geography courses as well as courses in other disciplines.

This chapter discusses the opportunities to integrate mapping and GIS, systematically and effectively, into curricula at the undergraduate level. It is geared for both the early career faculty member as well as the graduate teaching assistant who is curious to learn how spatial concepts may be taught and learned in the classroom. Our examples will illustrate how GIS and online mapping applications can be used by faculty who *do not* regularly use these tools in their teaching. Most importantly, we highlight the educational role that geographic technologies can play in higher education, in departments of geography, and beyond.

The long-term educational objectives of introducing spatial concepts and promoting spatial learning can be supported by mapping and GIS applications across a wide range of teaching and learning environments. These need not be hardware or software intensive, and can readily be scaled for use by those with little GIS background. At their most useful, these applications can be used in conjunction with other learning experiences and techniques—such as problem-based and field-based inquiries.

Fink (1999), a geographer and curriculum design specialist, strongly advocates for these active learning techniques, suggesting a three-step approach to their implementation: (1) expand the types of experiences instructors create, (2) take advantage of the "power of interaction," and (3) "create a dialect between experience and dialogue." The mindful use of digital mapping applications represents such an experience. As students manipulate, modify, query, analyze, and explore spatial data sets, they gain a richer understanding of the content, and the interactivity itself supports the learning process (Medyckyj-Scott 1994; Fink 1999). GIS can also be applied to problem-based learning experiences (Carver, Evans, and Kingston 2004; Drennon 2005) and building competencies and literacy in specific disciplines (Lo, Affolter, and Reeves 2002).

When GIS activities are well-planned and thoughtfully embedded into a curriculum, they can connect with and support constructivist-learning theory. Fink's taxonomy of significant learning includes such learning outcomes and processes as foundational knowledge, application, integration, human dimensions (students learning about themselves, as well as understanding and interacting with others), caring (developing values, interests, or feelings about a subject), and learning how to learn (Fink 2003, 27–59). By carefully linking GIS- and mapping-based exercises with curriculum of many different types and formats, the exercises can be used to advance learning in several of these areas. For example, viewing the tabular, descriptive attributes associated with geographic features at different places provides students with basic foundational knowledge of places. Creating and executing spatial models is an application that can help students develop critical, creative, and practical thinking skills as they modify parameters and evaluate results. The ability of GIS to merge and synchronize data from such a myriad of sources and scales is a form of Fink's integration. Through visual and analytical exploration, students cultivate a curiosity about the spatial data with which they are working. Curiosity leads to inquisitiveness, which in turn leads to learning how to learn. GIS alone does not equal learning, but it has the potential to support these forms of significant learning.

Visualization is another trajectory through which GIS supports teaching and learning. Visualizing the spatial relationships among data sets assists in the cognitive aspect of learning and promotes deeper understanding (MacEachren 1995; Kaiser and Wood 2001; Schultz 2004; Sinton and Lund 2007). The vast majority of geography students find that some form of visualization is helpful to the learning process (Alibrandi 2003; Schultz 2004). This is particularly true when the information would otherwise appear in a table or embedded within text, and also when the data are less tangible and more difficult to observe directly (such as some cultural patterns, the spatial distribution of attitudes or emotions, bathymetry, and groundwater contamination).

As an illustration, when we are teaching about the sources of political and cultural tension in Africa, it is helpful to understand the geographic relationship between the political boundaries of countries—representing a legacy of colonial power struggles—and cultural or ethnic boundaries, which are not likely to coincide. In Figure 9.1, we have used GIS to overlay the

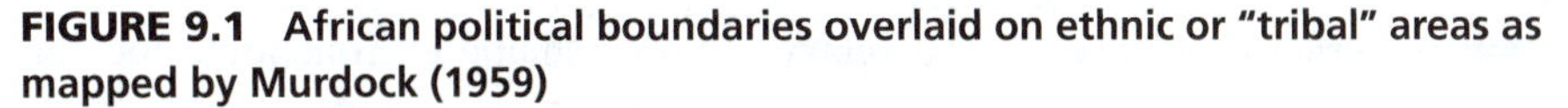
FIGURE 9.1 African political boundaries overlaid on ethnic or "tribal" areas as mapped by Murdock (1959)

Source: Adapted from a figure previously published by ESRI Press and used with permission.

African country borders over numerous and distinctive ethnic areas outlined by Murdock (1959) in *Africa: Its Peoples and Their Cultural History*. This simple overlay technique provides students with a visual understanding of one potential source of conflict in different regions of Africa. This could be helpful in any number of human geography or other social science courses.

As another example, an image depicting the locations of active volcanoes, plate tectonic boundaries, and continents is a highly valuable contribution to any class in which these geologic phenomena are discussed (Figure 9.2). By being able to visualize the spatial coincidence of these points, lines, and polygon features, students will gain a much deeper understanding of these fundamental relationships.

Finally, using a simple aerial image providing a birds-eye perspective of a place can help students better grasp its geography and patterns, versus having only a ground-based perspective. Google Earth or other virtual globes are ideal tools to help people appreciate the spatial distribution of buildings on a college campus, for example (Figure 9.3).

In September 2004, U.S. Secretary of Labor Elaine L. Chao announced a series of investments totaling nearly $6.4 million to address the workforce needs of the geospatial technology industry. The fact that such efforts are underway underscores the importance of introducing geospatial concepts into the curricula.

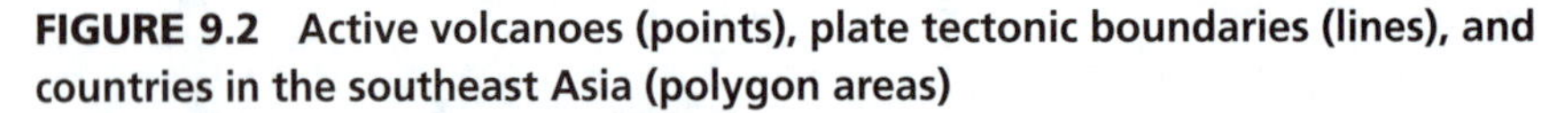

FIGURE 9.2 Active volcanoes (points), plate tectonic boundaries (lines), and countries in the southeast Asia (polygon areas)

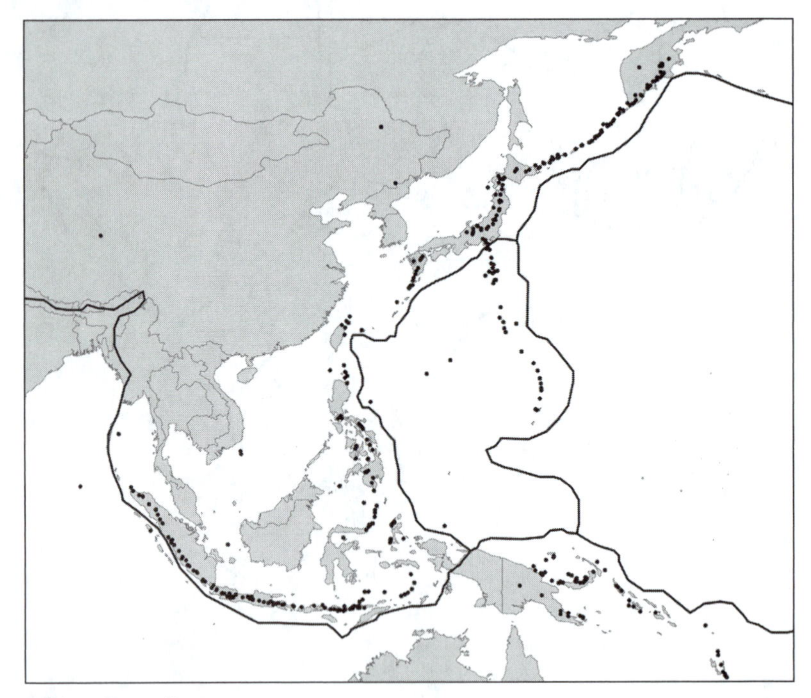

Source: Diana Stuart Sinton

FIGURE 9.3 Using Google Earth® to understand the spatial distribution of buildings on a college campus

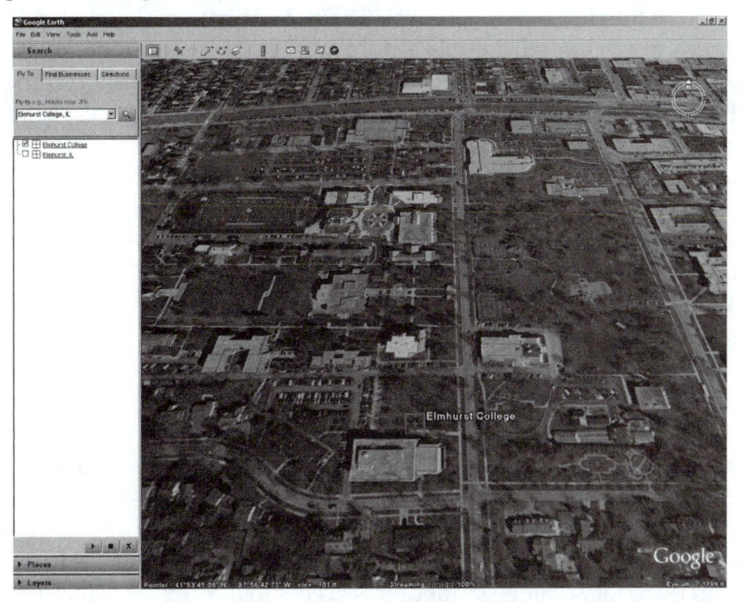

Source: Printed with permission from Google. Copyright 2007 Europa Technologies/Navteq. All rights reserved.

Since 2004, the U.S. Department of Labor (DOL) hosted forums with geospatial technology industry leaders, educators, and the public workforce system seeking to understand and implement industry-identified strategies to confront critical workforce shortages. DOL listened to employers, industry association representatives, and others associated with the geospatial technology industry regarding some of their efforts to identify challenges and implement effective workforce strategies. DOL's employment and training administration supports comprehensive business, education, and workforce development partnerships that have developed innovative approaches that address the workforce needs of business while also effectively helping workers find good jobs with good wages and promising career pathways in the geospatial technology industry.

Regardless of a student's intentions following his or her undergraduate education, we find that developing aptitude, expertise, competence, and confidence in all realms of spatial thinking and reasoning is a positive step forward (Hespanha, Goodchild, and Janelle 2006; Sinton and Lund 2007). In short, when spatial concepts are logically presented in the classroom, they address practical problems, promote critical-thinking and problem-solving skills, enable visualization of data, enhance patterns not normally seen without the use of geospatial tools, support quantitative reasoning, and challenge students to deeper levels of learning.

Davis (1993) notes that some students seem naturally enthusiastic about learning, but many need, or rather expect, their instructors to inspire, challenge, and stimulate them: "Effective learning in the classroom depends on the teacher's ability . . . to maintain the interest that brought students to the course in the first place" (Ericksen 1978, 1). Whatever level of motivation your students bring to the classroom will be transformed, for better or worse, by your interactions with them. Research has also shown that good teaching practices can do more to counter student apathy than special efforts to address motivational issues directly (Ericksen 1978). Most students respond positively to a well-organized course taught by an enthusiastic instructor who has a genuine interest in them and what they learn. Thus, activities one undertakes to promote learning will also enhance students' motivation. For GIS to play an effective role in those activities, instructors must be mindful of its potential and aware of its pitfalls. Without thoughtful planning into *how* one intends to use GIS or spatial tools as geographic teaching tools, students may never realize that potential or be any closer to learning spatial concepts (Kerski 2003; Sinton and Lund 2007). Our discussion of this will focus on two different models of GIS use: (1) Internet-based mapping applications and (2) creating a course- or project-specific spatial application for use by students.

The ease-of-use of many online mapping applications makes them logical choices for teaching purposes. No specialized software is required, and virtually every college student has access to a computer lab with a high-speed Internet connection (Table 9.1). For almost any geography course taught, there is an Internet map server that can provide related maps of interest. Virtual globes, such as the popular Google Earth, NASA's World Wind, and ESRI's ArcGIS Explorer, can be customized with a user's own data and perform basic GIS tasks within a three-dimensional environment. A third type of online mapping application allows users to create their own map "mash-ups"—converting a typical Google or Yahoo! street map into an environment where students can share stories over geographic spaces.

Integrating these types of tools into a learning environment can occur in multiple ways. Faculty can use the maps online during class to illustrate a point during a lecture, or assign students the task of reviewing a collection of maps (on their own time) before the next class session, in preparation for discussion. Maps can become the final product of a class, such as maps that explore cultural and socioeconomic characteristics of the college community (Trinity College 2007). This type of work could certainly complement an urban geography course.

Visualizing census data through maps is a common practice in many geographic fields. As one example, an economic geographer teaches a 100-level course entitled "Place and Society" in which students use census data to map areas of urban inequality. Students select a city to study and spend one lab session (2.5 hours) exploring thematic maps through the Census Bureau's American Fact Finder. During that one lab, the professor demonstrates how to use the application, and models for the students how the maps can be used to support their arguments. Students then have one

TABLE 9.1 Sample resources for introducing mapping, GIS, and spatial concepts in the classroom

Hardware-limited options

- Utilize traditional paper maps (*e.g.*, geologic, topographic, or planning maps) to introduce concepts of data-driven maps and spatial representations
- Assign homework exercises that access online resources from student-owned or campus computer labs

Software-limited options

- Generate, view, and/or print maps for exercises/labs from online sources, such as these examples (all Windows and Mac):
  - Census Bureau's American Fact Finder—http://factfinder.census.gov
  - Collections for Teaching and Research—http://ublib.buffalo.edu/libraries/asl/maps/researching_maps.html
  - David Rumsey's GIS browser—http://www.davidrumsey.com/GIS/
  - Environmental Protection Agency's EnviroMapper—http://maps.epa.gov/enviromapper
  - Gapminder—http://www.gapminder.org/
  - SocialExplorer—http://socialexplorer.com
  - United Nations GeoPortal—http://geodata.grid.unep.ch
  - United Nations Maps and Graphics—http://maps.grida.no/
  - U.S. National Atlas—http://www.nationalatlas.gov
  - WorldMapper—http://www.worldmapper.org
- Generate, view, and/or print map mashups using online tools, such as these (all Windows and Mac):
  - BatchGeocode—http://www.batchgeocode.com
  - Google's My Maps—http://maps.google.com
  - GPS Visualizer—http://www.gpsvisualizer.com
  - Platial—http://www.platial.com
  - Wayfaring—http://www.wayfaring.com/
  - Your G Maps—http://yourgmaps.com
  - Zee Maps—http://www.zeemaps.com
- Generate, view, and/or print maps for exercises and labs from GIS software. Some free or low-cost mapping applications include
  - Christine GIS—http://www.christine-gis.com/ (Windows only)
  - ArcExplorer JAVA Edition for Education (AEJEE, http://www.esri.com/software/arcexplorer/about/arcexplorer-education.html) (Windows and Mac)
  - Generic Mapping Tools—http://gmt.soest.hawaii.edu/ (Windows only)
  - MICRODEM—http://www.usna.edu/Users/oceano/pguth/website/microdem.htm (Windows only)
  - Manifold—http://www.manifold.net/ (Windows only)

(continued)

TABLE 9.1 Continued

- Generate, view, and/or print imagery or maps for exercises/labs from using online virtual globes, including:
  - Google Earth—http://earth.google.com (Windows and Mac)
  - NASA's World Wind—http://worldwind.arc.nasa.gov (currently Windows only)
  - ESRI's ArcGIS Explorer—http://www.esri.com/software/arcgis/explorer/index.html (Windows only)
  - EarthBrowser—http://www.earthbrowser.com/index.html (Windows and Mac)
  - Microsoft's Virtual Earth—http://maps.live.com/ (Windows and Mac)
- To use GPS in class or lab exercises, shareware can be used to download GPS data to computer (i.e. GPSVisualizer; EasyGPS; Garmin DNR; OziExplorer). Microsoft® Excel or other spreadsheet software can be used simply to plot data as (x,y) coordinates.

Some "ready-to-use" classroom exercises utilizing these ideas include

- SERC (Science Education Resource Center) collection of examples for "Teaching with GIS in the Geosciences"—http://serc.carleton.edu/introgeo/gis/index.html. Included is a searchable collection of examples—http://serc.carleton.edu/introgeo/gis/GIS_examples.html
- ESRI's ArcLessons—http://gis.esri.com/industries/education/arclessons/arclessons.cfm
- National Center for Case Study Teaching in Science,
  - Snowboarding in New York: A GIS Case Study—http://www.science cases.org/snowboard/snowboard_notes.asp
  - The Fate and Transport of Toxic Releases—http://www.sciencecases.org/tri_gis/tri_gis2.asp
- ESRI's Mapping Our World exercises—http://gis.esri.com/esripress/display/index.cfm

week to write a report (illustrated by five different maps) that addresses such questions as:

1. Are there areas where many groups of disadvantaged populations appear to be concentrated?
2. Are there groups with distinctly different distributions?
3. What factors might have contributed to the patterns your analysis reveals?
4. What are the implications of the patterns your analysis identifies?
5. Should policy measures be designed to address these differences, and if so what types of policy responses would you recommend?

A second model for integrating GIS into geography classrooms involves having faculty, teaching assistants (TAs), or students develop the maps themselves. Investing the time in this activity can pay large dividends as students gain exposure to working with spatial data by interacting with dynamic maps, understanding the information behind the images, and taking ownership of the maps by incorporating their own data. Again, thoughtful planning is necessary to ensure that the exercises promote learning of the discipline rather than excessive time spent learning the tool itself (Lloyd 2001).

Acquiring, purchasing, converting, or creating GIS-ready data sets, and then coordinating them to ensure they are ready to be used together, is by far the most time-consuming aspect of developing a GIS project. Fortunately, emerging interoperable functionality makes it increasingly easy to move data between different GIS software packages. Those who are uncertain how to build a mapping project from scratch should feel reassured that there are multiple possible approaches. With ESRI's ArcGIS software as an example, the list of different methods for bringing data into a GIS project could include:

- locating files of ArcGIS-ready data (shapefiles, geodatabases, coverages, grids, images) provided by ESRI or any other sources;
- joining data maintained in tabular form (such as an Excel spreadsheet or Access database) to a shapefile;
- displaying the (x, y) coordinates of event locations (from tabular data of latitude/longitude, or from Global Positioning System (GPS)), and then converting those to a point data set;
- geocoding tabular data (such as addresses), generating a point for each address;
- georeferencing a digital image of a map or photo;
- creating a new spatial data set by heads-up digitizing over an aerial photograph or a georeferenced map or image.

If a faculty member or TA spends the time to build a GIS project, and students are only expected to use or view the mapped layers (without creating any new data themselves), the students will experience a relatively easy learning curve. They can readily learn how to navigate around the graphical user interface, modify the display and symbology of data, and explore visual correlations. Builders of the GIS can decide whether students will access the data through simple and free or low-cost programs (such as ArcExplorer) or more complex, robust programs served through an Internet browser (such as a customized ArcIMS or ArcServer, MapServer, CartoGraph, or Maptitude's IMS). In these situations, the heavy work load rests on the faculty or TA to construct the GIS, but students are then exempt from having to learn software to access the maps (and can more effectively spend their time on the course content).

Alternatively, faculty may deliberately choose to have students build the GIS project themselves. Though the course content may have to be modified to accommodate the time that students will spend with mapping software, the process of having students participate in the creation can be very rewarding, moving them from passive to active learning. This parallels the role of fieldwork within the discipline of geography: We want our students to experience place and know it intimately, rather than rely on another's interpretation. The experience aspect solidifies the active learning process and presents an activity where deeper learning is encouraged. In any course where field data are collected (biogeography, historical geography, tourism geography, cultural geography, among others), the learning process could be enhanced further to have those data organized and displayed within a GIS.

As an example of this active learning model for using GIS, a professor of historical geography has students gather historical maps and images of an area. Each year this study area changes: It has been London, Boston, and even their own rural location in the United States. The historical maps can be acquired in paper format (in which case they are scanned during class) or digital format, and in each instance the students must georeference the maps within the GIS. Students also take digital images of buildings, parks, landscapes, and other locations of historical significance that they connect to their respective geographic locations within the GIS (via hyperlinks). The objective of the course activity is to describe how places in the city have changed with time and to suggest factors that might have initiated those changes. Their course-wide GIS is used by the students to both illustrate and support their arguments for why those changes might have occurred (measuring distances of places from a central business district or determining the relative density of available goods and services or calculating the proximity of public transportation lines to different socioeconomic groups).

Many students new to geography are unfamiliar with mapping concepts that professionals and experienced users may take for granted. The more exposure students have to seemingly simple geographic and cartographic concepts, the better prepared they are to understand more complex GIS tasks at a later stage. The introduction of hands-on map creation and interpretation exercises and the associated terminology can greatly enhance the learning experience of the students.

As mentioned earlier, there is often the misconception that a commitment to expensive hardware or software is necessary when introducing GIS and spatial concepts to students in the classroom. In fact, there are many options that may be pursued despite resource limitations or student difficulties with technology. A few ideas on what can be accomplished with different levels of resource availability or student background are listed in Table 9.1. The online activities for this chapter provide help and suggestions for using these resources to create learning materials for your classes. They include examples of how to use Google Earth and ESRI's ArcExplorer to developing value learning experiences.

In summary, the use of spatial data and maps is vitally important to many disciplines, not just to geography. The widespread availability of online mapping applications and relatively simple-to-use GIS software should mitigate any perceived hardware or software barriers that might

Teaching with Geographic Technologies

Cary Komoto

Newly emerging geographic technologies provide geographers with a bounty of options for teaching. There are a large number of questions that may be raised concerning the use of these technologies for student learning. Thoughtful planning regarding how GIS can be used as a geographic teaching tool is an example of scholarly teaching. With a few additional steps and a bit more structure, this idea can be transformed into a SoTL (scholarship of teaching and learning) project.

Unlike some of the SoTL examples in other chapters, the unique aspect of this scenario is the presence of a specific technology that can be used for various purposes. A starting point is to develop a more precise research question based on the use of spatial concepts. The researcher then needs to develop a research plan and begin a design phase. The key questions that should be addressed in the planning and design stage include (1) what previous studies have been done; (2) what evidence or data can or should be collected for analysis; (3) how to collect this evidence or data; (4) how to analyze the evidence or data; (5) what the final product of the research will look like; (6) how critical reflection will take place; (7) how will peer review occur; (8) and how the final product will be disseminated to the greater community. It is important to note that in SoTL the audience for the research will often help determine the methods, approach, and rigor of the study. The ultimate goal for much SoTL work is to inform others about student learning and how students learn.

Recommended Readings

Carver, S., A. Evans, and R. Kingston. 2004. Developing and testing an online tool for teaching GIS concepts applied to spatial decision-making. *Journal of Geography in Higher Education* 28(3):425–38.

Drennon, C. 2005. Teaching geographic information systems in a problem-based learning environment. *Journal of Geography in Higher Education* 29(3): 385–402.

Lee, J., and R. S. Bednarz. 2005. Spatial skills, cognitive maps, and map drawing strategies. *Journal of Geography* 104(5):211–21.

Wiegand, P. 2006. *Learning and teaching with maps*. Abingdon and New York: Routledge.

otherwise limit the use of dynamic maps in the classroom. Through these uses, we, as educators, have opportunities to promote active learning to our students as they gain confidence in their abilities to collect, manage, and analyze spatial data. We can use mapping and GIS to support the development of critical-thinking (and viewing) skills. In a world where globalization and spatial matters are becoming an essential part of one's knowledge base, the infusion of geospatial concepts into the curriculum will help our students succeed in a future world where space matters.

References

Alibrandi, M. 2003. *GIS in the classroom: Using geographic information systems in social studies and environmental studies*. Portsmouth, NH: Heinemann Publishers.

Carver, S., A. Evans, and R. Kingston. 2004. Developing and testing an online tool for teaching GIS concepts applied to spatial decision-making. *Journal of Geography in Higher Education* 28(3):425–38.

Davis, B. G. 1993. *Tools for teaching*. San Francisco, CA: Jossey-Bass.

Drennon, C. 2005. Teaching geographic information systems in a problem-based learning environment. *Journal of Geography in Higher Education* 29(3):385–402.

Environmental Systems Research Institute (ESRI). 2007. ArcGIS Explorer http://resources.esri.com/arcgisexplorer/ (last accessed 16 February 2007).

Ericksen, S. C. 1978. The lecture. Ann Arbor, MI: Center for Research on Teaching and Learning, University of Michigan, *Memo to the Faculty*, no. 60.

Fink, L. D. 1999. *Active learning*. http://www.ou.edu/idp/tips/ideas/model.html (last accessed 31 August 2007).

———. 2003. *Creating significant learning experiences*. San Francisco, CA: Jossey-Bass.

Hespanha, S. R., F. Goodchild, and D. Janelle. 2006. Spatial thinking and technologies in the undergraduate social science classroom. Society for American Archaeology (SAA) annual meeting held in San Juan, Puerto Rico 26–30 April 2006, unpublished paper.

Kaiser, W. L., and D. Wood. 2001. *Seeing through maps: The power of images to shape our world view*. Amherst, MA: ODT Inc.

Kerski, J. J. 2003. The implementation and effectiveness of geographic information systems technology and methods in secondary education. *Journal of Geography* 102(3):128–37.

Lloyd, W. J. 2001. Integrating GIS into the undergraduate learning environment. *Journal of Geography* 100(5):158–63.

Lo, C. P., J. M. Affolter, and T. C. Reeves. 2002. Building environmental literacy through participation in GIS and multimedia assisted field research. *Journal of Geography* 101(1):10–19.

MacEachren, A. M. 1995. *How maps work: Representation, visualization, and design*. New York: Guilford Press.

Medyckyj-Scott, D. 1994. Visualization and human-computer interaction in GIS. In *Visualization in geographical information systems*, eds. H. M. Hearnshaw and D. J. Unwin, 200–11. New York: John Wiley & Sons.

Murdock, G. P. 1959. *Africa: Its peoples and their cultural history*. New York: McGraw-Hill Publishers.

National Research Council. 2005. *Learning to think spatially: GIS as a support system in the K–12 curriculum*. Washington, DC: National Academies Press.

Schultz, R. B. 2004. A comparison of undergraduate geoscience course offerings: Online versus on-ground. *Abstracts with Programs*: Geological Society of America Annual Meeting, Denver, CO, Nov. 2004.

Sinton, D. S. and J. Lund. 2007. *Understanding place: Mapping and GIS across the curriculum*. Redlands, CA: ESRI Press.

Trinity College. 2007. FYFO232: Invisible Cities Google mash-ups. http://prog.trincoll.edu/gis/projects/fymashups/ (last accessed 30 August 2007).

CHAPTER 10

Looking Beyond the Lecture: Promoting Significant Learning in Large Classes

Douglas W. Gamble

Starting a new faculty position is an exciting time. All of the hard work has paid off and you have obtained a job that allows you to pursue your intellectual passion and curiosity. However, soon the challenges of the new position set in and you are faced with the task of balancing research, teaching, and service, as well as assimilating into a new university culture.

Teaching a large class is often one of these new challenges. Such a task is intimidating for many reasons. First, you probably have not been in a large class since your early undergraduate years so the experience is sometimes hard to remember. Second, you have probably become more accustomed to the small classes that predominate in graduate school. Another reason for the intimidating nature of the large class assignment is the prevailing attitude in academia that large classes are not the ideal learning environment. It is assumed by many academics that a large class leads to poor student performance, student discontent with university education, and instructor frustration with the educational process (McKeachie 1980; Chism 1989; Michaelsen 2002). Consequently, many aspiring academics face the challenge with a degree of dread and follow the path of least resistance by lecturing and giving multiple-choice tests. Gardiner (1994) interviewed 1,800 college teachers and found that roughly three-quarters of the instructors identified lecture as their usual instructional strategy. Such a strategy for teaching a large class does not necessarily equate to poor teaching and learning, but it is not always the most appropriate format for engaging students and creating significant learning experiences for them and a positive teaching experience for you.

Despite the limitations of the large class format in higher education, there has been a successful movement to expand the traditional lecture format of large classes and engage students through active learning. Such efforts have resulted in a greater degree of satisfaction for both the students and the instructor, as well as greater student performance, retention of knowledge, critical thinking, and attitude change (MacGregor et al. 2000). The purpose of this chapter is to outline how you can develop a successful instructional strategy for large classes that moves beyond the traditional lecture format and actively engages students in significant learning. Such active, engaged learning is accomplished in a large class through appropriate course design and alignment that utilizes active learning strategies, as opposed to simply peppering a lecture with the occasional active learning exercise and technique. Using Activity 10.1 on this book's web site, you can put these ideas into action in your own courses.

Putting Your Course Online

Kenneth Foote and Michael Solem

Today's software tools make it very easy to create hypertext teaching materials, but knowing the basics of web-authoring software does not automatically result in effective courseware. Faculty often need very practical advice about how to design and plan their projects, how to get the most out of the time they will invest in authoring, and how to avoid common pitfalls of hypertext design. Preparing courseware on the web should result not only in *new* materials but also in *better* materials. Unfortunately, faculty do not always have time to explore, for example, the extensive literatures of curriculum development, graphic design, and evaluation and assessment. They need to-the-point, user-friendly tips that can get them off the ground quickly.

For these reasons we have included sidebars throughout this chapter so you can begin to think of how web media can further engage your students and enhance your lectures in classes large or small. The tips we share have been used successfully with geographers who participated in the Virtual Geography Department workshops at the University of Texas from 1996 to 1998 and in faculty workshops at the University of Colorado since 2000. Though technology can change rapidly, the underlying principles of effective instruction are much more stable. No matter what the technology, the fundamental question you need to ask when setting course objectives is: What will students gain from working with technology?

It should be noted that I do not believe that lecturing is inappropriate for teaching a large class; indeed, lecturing can be a very effective component of teaching a large class as noted by many researchers (Costin 1972; Agnew and Elton 1998; Cuseo 1998; Johnson, Johnson, and Smith 1998; McKeachie 1999; Bain 2004). I believe that an instructor needs to recognize the learning opportunities for which lecturing is appropriate, and those opportunities for which other instructional formats are appropriate. Bligh (2000) indicates that lectures are just as effective as any other method for transmitting information, but they are not as effective as discussion in promoting student thought, nor should they be expected to change student attitudes or behavioral skills. Thus, if one goal of your course is to convey information, the lecture is very appropriate for portions of the class.

THE LARGE CLASS IN HIGHER EDUCATION

Fundamental to this discussion is the question: what exactly is a large class and what role does it currently play in higher education? There is no one accepted definition of a large class (Weimer 1987). A commonly used threshold value is fifty or more students, and this size will be used as the large class designation for this chapter (MacGregor et al. 2000; Stanley and Porter 2002). No matter the definition of a large class, there is no doubt that large classes are becoming more and more prevalent at colleges and universities because of decreased funding and increased enrollment (McKeachie 1980; Zietz and Cochran 1997). Further, the majority of the instruction of large classes at universities is conducted by early career faculty, graduate students, and adjunct or part-time instructors, rather than by experienced faculty (Stanley and Porter 2002).

The main reason for large classes is to lower the overall cost of a college education through the efficient use of faculty resources, and such classes allow for effective standardization of course material (Chism 1989). However, large classes can create negative student attitudes and inhibit learning through student anonymity and diminished social interaction (Marsh, Overall, and Kesler 1979; Feldman 1984; Williams et al. 1985; Gilbert 1995; McKeachie 1999; Michaelsen 2002). The result is a course with an impersonal nature, a limited range of instructional activities, low motivation for students, and that stresses rote memory of facts as compared to development of critical thinking (Siegel, Adams, and Macomber 1960; Chism 1989).

Given such negative factors, is it possible to create a significant learning experience in a large classroom? I believe so, and I believe that you can create a significant learning experience in your large class through careful alignment of day-to-day activities and course goals, emphasis upon active learning, and looking beyond the traditional lecture format. Such efforts can be facilitated by following the advice of Fink (2003): think about student learning as opposed to *your* teaching. Once you focus on student learning, the next obstacle to overcome is the limitations imposed by the situational factors of the large class. Such situational factors include, but are not limited to, classroom layout, the number of students, class material delivery (live and online), web resources, and available teaching assistants (TAs). Fink (2003) identifies dealing with situational factors as the first step in integrated course design. Thus, you should recognize the specific context of the large lecture format, addressing the limitations and devising techniques that take advantage of its unique features.

Why Use the Web?

Kenneth Foote and Michael Solem

Virtually every college and university in the United States and many other nations have, over the past several years, developed training courses or workshops to show their faculty how to use the web in their courses and how to design online instructional materials. Many campuses have developed their own materials for these courses, including authoring toolkits. Other institutions offer their faculty licensed access to e-learning course platforms such as Blackboard, which provides ready-to-use templates and a suite of online communication tools that help faculty get started. These systems also provide tools for course management. But, for the most part, these training programs only cover the basics of hypertext authoring and barely scratch the surface of the issues underlying web design and the creation of effective hypertext materials.

The literature is stocked with articles, essays, and commentaries about educational applications of web technology, but for the sake of brevity we will cite four reasons by Rudenstine (1997) that, although written when the web was first emerging in higher education, have not lost their relevance:

1. *The Internet can provide access to essentially unlimited sources of information not conveniently obtainable through other means.* Popular examples for geographers include the online databases of the U.S. Census Bureau (www.census.gov) and the National Oceanic and Atmospheric Administration (www.noaa.gov).
2. *The Internet allows for the creation of unusually rich course materials.* Web pages can support video clips, audio files, full-color digital photos and maps, animated slides, and networks of links to related information sources, providing the instructor with an unprecedented palette for authoring course materials.
3. *The Internet enhances the vital process of "conversational" learning.* The capabilities of electronic communication tools such as discussion boards, wikis, blogs, and chat rooms mean that course-related interactions can take place in an "anytime, anywhere" fashion.
4. *The Internet reinforces the conception of students as active agents in the process of learning, not as passive recipients of knowledge from teachers and authoritative texts.* Not only does web-authoring software and publishing tools make it easy for faculty to put their course online, but these resources also enable

students to contribute directly in developing materials that can build community-based course web site showcasing examples of student work.

Recommended Readings

Blackboard, Inc. 2007. Blackboard. http://blackboard.com/us/index.Bb (last accessed 16 Feburary 2007)

Horton, S. 2000. Web teaching guide: A practical approach to creating course web sites. New Haven: Yale University Press.

Rudenstine, N. L. 1997. Point of View: The Internet and Education, A Close Fit. Chronicle of Higher Education 21 February, A48.

KEY CONCEPTS FOR SUCCESSFUL LARGE CLASSES

I would like to highlight six key concepts I have found particularly useful during my fifteen years of experience teaching large classes. These relate to the larger body of literature that covers these and other strategies in more detail (Biggs 1999; MacGregor et al. 2000; Stanley and Porter 2002; Fink 2003; McKeachie and Svinicki 2006).

Set Clear Goals and Help Students Understand Them

From my experience, the majority of lower-level courses serve multiple purposes such as introduction to the major, a required class for a major, or a university-wide requirement. The key to the successful instruction of a large class is addressing multiple course goals by moving beyond the lecture and focusing upon active learning. However, before jumping directly into the nuts and bolts of integrating active learning into your course, you must develop clearly stated learning goals to which you link classroom activities. Fink (2003) provides excellent guidance in the development of such goals, and I encourage the reader to consult his material.

The construction of clear goals is only one-half of the effort that should be put into this task. You should also review the goals and link them to course material in your instructional efforts as often as possible. For instance, in my classes I spend time on the first day of class reviewing the course goals in detail and then each time I finish a section of the course or review for a test, I review the goals again so that students have a clear understanding of what I am attempting to accomplish. Such a revisiting of the course goals is particularly effective when your assessment or tests are closely aligned with the goals because it offers a study strategy for the student. For example, if one goal of the course is to introduce the student to terminology used in the field, the student will know to review the vocabulary for tests. Once you have constructed clearly stated goals and devised a strategy to develop student understanding of the goals, it is time to focus upon active learning.

Structure a Course for Use of Active Pedagogy to Promote Learning

Active learning has received much attention in the last ten years as an effective strategy in increasing the quality of large classes (MacGregor et al. 2000; Buckley et al. 2004). Many definitions exist, but I particularly like Gold et al.'s (1991): Active learning is achieved when a student moves beyond passive reception of knowledge and explores existing knowledge to create their own knowledge. Successful implementation of active learning requires an alignment of course goals and activities to synergistically achieve student engagement. In Fink's (2003) holistic model of active learning, this synergy is created through providing information or ideas, a call for students to do or observe something, and prompting students to reflect on their learning. You can find many good guides to course development and some are particularly relevant to large classes. Perhaps one of the best is provided by Porter and Stanley (2002) and I encourage the reader to consult this resource. Their advice on how to develop an effective large class can be condensed into the following steps: plan early, manage time well, seek advice from experienced large class teachers, get to know your students, attend to classroom management, use active learning strategies, use technology appropriately, develop effective testing mechanisms and grading procedures, select and train TAs and support staff, teach for inclusion, and remember that large classes provide both teaching and learning challenges.

All of the issues listed by Porter and Stanley (2002) are important and allow for the development of an effective large class. However, it is quite an intimidating list and realistically cannot be completed in an instructor's first semester given other professional obligations. So I have found that as one starts to develop a large class, it is essential at first to stress three concepts: course structure, active learning techniques, and classroom management. By focusing upon these three issues, a new instructor can lay the foundation for a successful large class and integrate other issues as outlined by Porter and Stanley into the course as it continues to develop. I have also included a template that you can use to address these concepts and to efficiently develop an active learning-focused large class in Activity 10.1 on this book's website.

To effectively design an active learning-focused course structure, do not schedule lectures and tests as the only class activities. Plan for additional active learning-based techniques. The existence of an accompanying lab or discussion section is a common strategy to integrate active learning into a large class (McKeachie and Svinicki 2006). However, this is not the only means of facilitating active learning. Active learning techniques can be integrated into the traditional lecture hall time so that they combine with the lecture component to create a more meaningful learning experience.

I have developed a large class focused on active learning (GGY 230 Introduction to Weather and Climate, typical enrollment 250–300 students, no accompanying laboratory sections) that utilizes both lecture and activities within two 1.25-hour lecture periods a week (Tuesday and Thursday) across a fifteen-week semester. This structure entails a lecture on Tuesdays and a combination of lecture and active learning activities on Thursdays. I have decided upon retaining the traditional lecture component of the course because one of the goals of the class is student acquisition of fundamental atmospheric science knowledge. Lecturing helps me obtain this goal.

On "active learning" days, I may finish the lecture from the previous period, but rarely do I lecture for more than twenty minutes. After this lecture, the remainder of the period is spent on active learning exercises completed by either individual students or small groups of students. I find that this format does not create a substantial loss of material covered in the course because it focuses my lectures and eliminates tangential ramblings. More importantly, the activity creates an opportunity for the student to experience the entire active learning process: information and ideas, experience, and reflection.

Use a Variety of Active Learning Techniques

It is important to use a variety of active learning techniques in your courses because the variety stimulates and retains student interest and addresses different student learning styles. Many sources exist that list a wide range of active learning techniques. I recommend Angelo and Cross (1993), Davis (1993), Finkel (2000), McKeachie and Svinicki (2006), and Geography Faculty Development Alliance (2007). I will provide four examples of active learning you can use in a large class format. In addition, Table 10.1 provides a list and brief description of potential active learning exercises for the large class format.

The first example of active learning I use is a combination of individual exercises and group discussion of exercises. In this technique, each student is first required to complete the exercise or question, and then a group of five to ten students chooses which solution in their group is the best (see the example of an active learning activity on this book's web site). At the end of the class, the students place each of their solutions and the identified best solution in a group folder. For assessment, each student is given credit for creating a solution (a one or zero) and a grade (on a scale of one to ten) is assigned to the chosen group response. Such group choice of best response reduces the total number of written responses that have to be graded by the instructor but also allows for some type of individual assessment (credit) for completing the task.

By creating small learning groups, I have found students tend to be more engaged in class. I have observed that group work allows for the development of critical thinking and communication skills within the context of weather and climate (a second goal of the class). Also, I have a higher degree of interaction with students because I walk around during group discussion, listening

TABLE 10.1 Potential active learning exercises for the large class format

Encouraging student questioning: pause during a lecture to ask students if they have any questions, allow questions throughout the entire lecture, or save time at end of class for questions.

Student-led discussions: assign students to lead small-group discussion of test material or a controversial topic related to course material.

Learning cells: cooperative formation of learning pairs in which students alternate asking and answering questions, also called student dyad.

Study groups: students form small groups of classmates with whom they study course material outside of class.

Buzz groups: form small groups to briefly discuss a problem derived from course material.

Problem posting: instructor lists student problems or questions concerning the course and addresses each one.

Two-column group discussions: list opposing arguments surrounding an issue in "for" and "against" columns on the board. Both faculty and students then discuss the merit of each for and against argument.

Student comparison of notes: pause lecture for students to compare notes and then ask for clarification or missing information.

Fishbowls: a class within a class in which a small group of students perform a task and others observe, also called inner circle.

Interviews: students interview other students or people outside of class about class-related material.

Debates: students lead a formal discussion of the merits of opposing sides of a controversial issue.

Case studies: small-group solving of a problem with several alternative solutions, many times represents a previous real-world situation.

Problem-based learning: provide a problem to students before the study of a subject to stress and stimulate their problem-solving skills.

Service learning: significant learning experience created by completion of a relevant, meaningful service activity during course work.

Think–pair–share: request that students reflect on subject or problem and then share with another classmate.

Pyramid activities: a student compares his or her work with one other person, then the students compare their completed work with another pair, then the four students discuss the work with another four students.

and talking with the students. Doing so helps me identify concepts that students are having difficulty understanding, and it also contributes to a feeling of classroom community. It is important with individual and small group activities to provide solutions to the exercise right after the group work has been completed (Michaelson 2002). In my class, I collect group folders and then discuss an answer key or problem solutions with the class as a whole.

One aspect of such small group activities that many colleagues have inquired about is nonparticipation of students in group work. I have found the combination of individual and group exercises offers two ways to assess participation. First, each student shares an individual assignment with the other members of the group. If a student is not present and participating, then he or she does not receive credit for the individual assignment or chosen group solution. Second, at the end of the semester each student is asked to assess, on a scale of one to ten, each group member in terms of quality of participation. The responses are averaged for each student and this average score is the basis for 5 percent of his or her overall grade. Overall, for the final semester grade, the weight given to activities is 5 percent for group work, 5 percent for individual work, 5 percent for peer evaluation, and 85 percent for four exams (see syllabus on the book's web site).

Another effective active learning exercise that my colleagues in human geography use with particular effectiveness is role playing. In such an activity, the instructor and TAs stand at the entry to the class and as students walk in, instruct them to sit in a specific location based upon a physical characteristic such as hair color, eye color, height, or sex. The result is a class segregated by a particular physical feature. Then the instructor chooses one particular group of students and asks them to make a decision for one of the other groups of students. Such a decision could be phrased as, "Students with blonde hair, should the students with brown hair stand the entire class or will you allow them to sit?" Obviously, the students will not be forced to stand the entire period. Rather after the students with blonde hair make a decision, ask each student to write a couple of sentences on how they felt about the segregation and decision. After the students have written their response, ask for volunteers or ask specific students to share what they have written with the class. A combination of this exercise with a lecture on political hegemony, colonialism, or racial segregation can create an active learning lesson that addresses the three active learning components of information and ideas, experience, and reflection.

A third example of active learning that I integrate into my class is the one-minute paper. A one-minute paper is exactly what it seems, students write about a specific topic or question for one minute and then stop. I typically have the students write a one-minute paper after they have watched a videotape that provides both sides of the global warming or coastal development debate. The students then write a one-minute paper explaining with which side of the argument they agree. When the papers are finished, students pass them to the person sitting next to them. This person reads the one-minute paper and then asks the writer for any clarifications. I then choose five to seven students to share with the rest of the class what they liked the most about the one-minute paper they just read. After the discussion, all one-minute papers are collected and each student is given credit for completing the activity.

I found the last example of an active learning exercise in an excellent resource by Angelo and Cross (1993). This book focuses upon classroom assessment in its broadest sense: obtaining useful feedback on what, how much, and how well students are learning. This exercise was originally designed for a class in cultural anthropology but through some simple modification I found it just as relevant to my "Introduction to Weather and Climate" class. The exercise is designed to assess students' prior knowledge of the course material, a particularly important issue when teaching a course that fulfills a general education requirement for nonmajors.

The exercise entails the instructor sharing a list of key terms for the course with students. For my class these terms include greenhouse effect, Hurricane Katrina, humidity, figures of the global circulation model, El Niño-Southern Oscillation, and the earth's radiation budget. For a cultural anthropology class the terms include the Weimar Republic, Senator Joseph McCarthy, and the Golden Triangle. The students then choose their level of background knowledge of each term from the following choices:

a. Have not heard of this
b. Have heard of it but do not really know what it means
c. Have some idea what this means, but not too clear
d. Have a clear idea what this means and can explain it

After students have completed the "background knowledge probe," the instructor discusses and links each term to a specific course goal. I also have the students complete the probe with computer-gradable Scantron sheets and share the results with them in the next class. Beyond providing essential information to the instructor, the exercise sets the tone of the class with students experiencing and reflecting, and the instructor providing information and ideas. Such a start to a course is cited by Fink (2003) as a requirement for developing a significant learning experience.

Classroom Management and Setting Standards for Student Conduct

In order to effectively teach a large class, one must take classroom management seriously. I guarantee that if you do not dutifully manage a large class with an active learning component it will be a disaster. Classroom management includes but is not limited to timely development of classroom materials, keeping class activities on time, handing out materials, keeping noise level down, collecting materials, training and guiding TAs, and addressing individual concerns. The keys to effective management are clarity, preparation, and flexibility. This way you can constantly refer to the document when confusion erupts. If chaos does ensue, remain calm and address the confusion as quickly and clearly as possible. Posting clarifications on the course web page is particularly effective.

Designing Your Pages and Web Sites

Kenneth Foote and Michael Solem

As you begin to author your web materials, there are several issues you should keep in mind. These include the following:

Audience and Access

1. What is the overall level of study and research skills among students in your class? How experienced are they with the Internet and web?
2. How will students access materials (*e.g.*, from home, during class, or both)?
3. What sort of classroom and laboratory facilities are available for students with otherwise limited access to the web?
4. How much help will students need? Is a "help staff" available? You may need to create exercises to help students become critical, knowledgeable users.

Getting Started and Setting Priorities

1. Begin by converting some existing paper materials, then adapt new ideas. Start with materials you know best or want to feature in your site.
2. Gain experience with the new media and consider its possibilities, but there is no need to start with the "flashiest" tools. Aim for substance.
3. Rethink and redevelop materials as they are put online, but always remember to ask yourself: What do students need or use most?

Be Aware of Copyright Law and Permissions

1. Materials that can be used in class under "fair use" cannot be republished on the web.
2. A good deal of copyright free material is available on the web.
3. Many people are willing to share their materials. Always ask before you borrow or use material from other web sites.

Employ Technology Thoughtfully to Enrich, Not Overwhelm, the Learning Environment

I use PowerPoint to deliver my lecture and I find it and other technologies to be effective in a large lecture format. However, be careful when using technology because it is important that you use it to enhance student learning, as opposed to making your job easier. Davis (1993) provides an excellent series of chapters that include practical tips on how to use a wide variety of technologies in the classroom. The greatest advantage I see with new technology is efficient classroom management. Through programs such as WebCT, or the equivalent software on your campus, previously onerous tasks such as posting grades, announcements, handing out class materials, mass e-mailings, and administering tests are much easier and more efficient.

In terms of the negatives of technology, Tufte makes a compelling argument that PowerPoint is "presenter-oriented, not content-oriented, not audience-oriented," and if not properly used the core ideas of teaching "are contrary to the cognitive style of PowerPoint" (2006:4,7). Consequently, we may not be aware of it, but PowerPoint may be causing our lectures to be more about teaching efficiency and less about student learning. The point here is not to avoid using PowerPoint, or other technology such as the Internet, geographic information systems, mapping programs, or Google Earth. Rather the point is to use technology to advance student learning as opposed to instructor convenience.

Save Time for Silence, Reflection, and Questions

A class of over 100 students may seem contradictory to the concept of silent reflection. However, in order for active learning to be effective, students need to reflect on the material and it is possible to facilitate such reflection in a large lecture hall. One technique that I find helps encourage reflection is a short break at the middle of a lecture period that allows students to review their notes. Such a break allows students to reflect in class, and I have found that classes remain very quiet. After the break, students are encouraged to ask for clarifications on the lecture material. In addition I finish lectures or activities five minutes before the official end time of the class to allow time for individual questions. You will be amazed at how willingly students talk to you when they are not rushing off to the next class or have to find your office. When students realize there is time set aside for interaction with the professor every class period, usually there is less confusion and less time spent answering repeat questions. This individual question time more than makes up for the three or four minutes I could have spent squeezing in one more piece of information during lecture.

CONCLUSION

You must recognize that teaching a large class is frequently the reality of a new position and can be a key to positive professional development. Successful instruction of a large class involves looking beyond the lecture and engaging a class through active learning that emphasizes information and ideas, experience, and reflection. In the development of an active learning-centered large class, it is important initially to focus on course format, ensure that your activities are aligned with course objectives and assessment methods, and attend to classroom management. And remember, development of an effective large class is not a process that is normally completed in one semester. Improving your teaching ability can be a career-long process and development of an effective large class teaching strategy is likely to take several semesters of tinkering.

Engaged Learning in Large Classes

Cary Komoto

This chapter provides a large number of possible questions that provide the starting point for a scholarship of teaching and learning (SoTL) project. For example, how can active learning enhance student performance and interest in large classes? One place to start with such a project is to define the learning goal for the active learning activity. This should be followed by a literature review and the development of a methodology for evaluating how well the activity meets the relevant learning goal. In this instance, a variety of assessments might be used, ranging from more traditional exams to less conventional assessments such as student interviews. The key here is to clearly articulate the learning goal and determine the best way(s) in which to substantiate that the goal is being met.

It might be useful to take the learning goal and express it in terms of student performance, which can provide a specific way of collecting evidence that either supports the contention that active learning is better than some other activity, or is not as effective. In this example, when designing the research one must consider how the evidence will be collected from students. This might include regularly scheduled exams or assignments, or it might require classroom observation. Whatever form of evidence is desired, it must be planned out before data collection can begin. In addition, for ethical reasons, researchers must seek approval from their institutional review board before conducting any research involving students as human subjects. Once data collection has been completed, a plan for analysis should already be in place. Whether this is statistical analysis or qualitative analysis, the analysis should be appropriate to the research question and the evidence collected.

After analysis is completed and conclusions are written, the traditional way to obtain peer review and dissemination is through the publication process. In a peer-reviewed journal, manuscripts are sent to reviewers with expertise in the topic to ensure that there were no ambiguities with methods, analysis, or conclusions. If deemed acceptable by the journal editor, the work can be disseminated through the publication of the article, and its findings become part of an established body of knowledge.

Recommended Readings

Bligh, D. A. 2000. *What's the use of lectures?* San Francisco, CA: Jossey-Bass.

Healey, M., and J. Roberts. 2004. *Engaging students in active learning: Case studies in geography, environment, and related disciplines.* Cheltenham, U.K.: University of Gloucestershire, Geography Discipline Network.

References

Agnew, C., and L. Elton. 1998. *Lecturing in geography.* Cheltenham, U.K.: Geography Discipline Network, University of Gloucestershire.

Angelo, T. A., and K. P. Cross. 1993. *Classroom assessment techniques: A handbook for college teachers.* San Francisco, CA: Jossey-Bass.

Bain, K. 2004. *What the best college teachers do.* Cambridge, MA: Harvard University Press.

Biggs, J. 1999. *Teaching for quality learning at university: What the student does.* Buckingham, U.K.: Open University Press.

Bligh, D. A. 2000. *What's the use of lectures?* San Francisco, CA: Jossey-Bass.

Buckley, G. L., N. R. Bain, A. M. Luginbuhl, and M. L. Dyer. 2004. Adding an "active learning" component to a large lecture course. *Journal of Geography* 103: 321–37.

Chism, N. V. N. 1989. Large enrollment classes: Necessary evil or not necessarily evil? *Notes on Teaching* 5: 1–8. Columbus, OH: Center for Teaching Excellence, Ohio State University.

Costin, F. 1972. Lecturing versus other methods of teaching: A review of research. *British Journal of Educational Technology* 3(1): 4–20.

Cuseo, J. 1998. Lectures: Their place and purpose. *Cooperative Learning and College Teaching* 9(1): 2.

Davis, B. G. 1993. *Tools for teaching.* San Francisco: Jossey-Bass.

Feldman, K. A. 1984. Class size and college students' evaluations of teachers and course: A closer look. *Research in Higher Education* 21(1): 45–116.

Fink, L. D. 2003. *Creating significant learning experiences.* San Francisco: Jossey-Bass.

Finkel, D. L. 2000. *Teaching with your mouth shut.* Portsmouth, NH: Boynton/Cook Publishers, Heinemann.

Gardiner, L. F. 1994. Redesigning higher education: Producing dramatic gains in student learning. *ASHE-ERIC Higher Education Report No. 7.* Washington, DC: George Washington University.

Geography Faculty Development Alliance. 2007. http://www.colorado.edu/geography/gfda/gfda.html (last accessed 31 July 2007).

Gilbert, S. 1995. Quality education: Does class size matter? *CSSHE Professional Profile* 14: 1–6.

Gold, J. R., A. Jenkins, R. Lee, J. Monk, J. Riley, I. D. H. Shepherd, and D. J. Unwin. 1991. *Teaching geography in higher education: A manual of good practice.* Oxford, U.K.: Basil Blackwell.

Johnson, D. W., R. T. Johnson, and K. A. Smith. 1998. *Active learning: Cooperation in the college classroom,* 2nd ed. Edina, MN: Interaction Books.

MacGregor, J., J. L. Cooper, K. A. Smith, and P. Robinson. 2000. Editors' notes. *New Directions for Teaching and Learning* 81(Spring): 1–4.

Marsh, H. W., J. U. Overall, and S. P. Kesler. 1979. Class size, students' evaluations, and instructional effectiveness. *American Educational Research Journal* 16(1): 57–69.

McKeachie, W. J. 1980. Class size, large classes, and multiple sections. *Academe* February 24–27.

———. 1999. *Teaching tips: Strategies, research and theory for college and university teachers,* 10th ed. Boston: Houghton Mifflin.

McKeachie, W. J., and M. Svinicki. 2006. *Teaching tips: Strategies, research and theory for college and university teachers,* 12th ed. Boston: Houghton Mifflin.

Michaelsen, L. K. 2002. Team learning in large classes. In *Engaging large classes: Strategies and techniques for college faculty*, eds. C. A. Stanley and M. E. Porter, 67–83. Bolton, MA: Anker.

Porter, M. E., and C. A. Stanley. 2002. Summary of key concepts for teaching large classes. In *Engaging large classes: Strategies and techniques for college faculty*, eds. C. A. Stanley and M. E. Porter, 324–29. Bolton, MA: Anker.

Siegel, L., J. F. Adams, and F. G. Macomber. 1960. Retention of subject matter as a function of large-group instructional procedures. *Journal of Educational Psychology* 51: 9–13.

Stanley, C. A., and M. E. Porter. 2002. Teaching large classes: A brief review of the research. In *Engaging large classes: Strategies and techniques for college faculty*, eds. C. A. Stanley and M. E. Porter, 143–53. Bolton, MA: Anker.

Tufte, E. R. 2006. *The cognitive style of PowerPoint: Pitching out corrupts within.* Cheshire, CT: Graphics Press.

Weimer, M. G. 1987. *Teaching large classes well.* San Francisco: Jossey-Bass.

Williams, D. D., P. F. Cook, B. Quinn, and R. P. Jensen. 1985. University class size: Is smaller better? *Research in Higher Education* 23(3): 307–17.

Zietz, J. and H. H. Cochran 1997. Containing costs without sacrificing achievement: Some evidence from college-level economics classes. *Journal of Education Finance* 23: 177–92.

CHAPTER 11

Teaching in the Field

Jennifer Speights-Binet and Douglas W. Gamble

One of the most rewarding and enjoyable aspects of being a geographer is spending time in the field. In fact, being in the field, that is the love of traveling, exploring new places and landscapes, and learning new ideas outside the traditional setting of the classroom, is what compelled many of us to pursue geography at the graduate and professional level. When we finally reach the point where we can teach our own geography courses, designing and implementing field experiences is both an exciting and overwhelming prospect. When solicited for advice, many veteran professors who have been teaching in the field for years will respond that you just have to get busy and do it. While there is something to be said for simply getting started, this jump-in-blind approach inevitably results in unsatisfying experiences for both students and instructors. The hard truth is field experiences (both short-term and long-term) require careful preparation, both in and out of the classroom, if they are going to be successful and geographically meaningful. Thankfully, some geographers who are committed to experiential, field-based learning have published articles offering useful tips for successfully teaching in the field. This chapter offers a brief summary of this helpful literature, while also discussing some of the authors' experiences and lessons learned. While considering field exercises in general, we will also address specific challenges associated with fieldwork focusing on human and physical geography content.

FIELD STUDY AND SIGNIFICANT LEARNING

Field study has long been recognized as a core activity for geographers (DeLyser and Starrs 2001). Sauer (1956: 296) states that "the principal training of the geographer should come, wherever possible, by doing fieldwork."

Despite the recognized importance of fieldwork, many have claimed that efforts in field instruction are decreasing in today's higher education system (Lounsbury and Aldrich 1986; Rundstrom and Kenzer 1989; Haigh and Gold 1993; Clark 1996; Higgitt 1996; Kent, Gilbertson, and Hunt 1997; Nairn, Higgitt, and Vanneste 2000; Salter 2001). Some of the reasons cited for this decline are tight university budgets, institutional focus on research, safety and insurance concerns, and a new emphasis on geographic information systems (GIS) analysis. One result of this decline in geography field instruction is a limited number of opportunities to acquire formal training in the development of field instruction techniques. Consequently, new academics are often left with the task of cobbling together an effective field instruction strategy with limited resources, little supervision, and the pressure to teach well the first time. Such a combination of limitations and pressures can discourage utilizing fieldwork in instructional activities. Unfortunately, this eliminates a potential teaching strategy that can increase the overall effectiveness and quality of your teaching efforts.

Given this current de-emphasis of field instruction and obstacles to developing such instructional activities, one may ask: why should I use field instruction? The main reason is that field instruction can increase the overall effectiveness of your teaching efforts. A well-designed field exercise can address a variety of learning outcomes. Through careful alignment of field instruction with course goals, all six of Fink's interactive taxonomy of significant learning (foundational knowledge, application, integration, human dimensions, caring, and learning how to learn) can be addressed (Fink 2003). In terms of foundational knowledge, field study provides basic understanding of a landscape, a fundamental concept in geography. Students achieve application of knowledge by solving problems with data collected during field exercises. Through field study, students integrate concepts and reach a true understanding of the physical and human processes that create landscapes. Further, learning in the field places students in real-world environments settings that can test them and their ability to interact with others. Students frequently cite field instruction as one of the most valuable learning experiences in their education, creating a caring component to learning. Lastly, in the completion of fieldwork, forces beyond our control often make students adapt in order to complete a task or reach a goal. In such cases, they learn the value of flexibility and the process of trial and error. In short, they are learning to learn.

Beyond addressing multiple learning outcomes, field instruction also offers an excellent opportunity to integrate active learning into instructional efforts. Active learning is a concept that has received much attention over the last ten years as a particularly effective strategy for increasing educational quality (Buckley et al. 2004). Gold et al. (1991) describes active learning as when a student progresses beyond receiving knowledge and explores existing knowledge to create personal knowledge. The advantages of field instruction as an active learning technique are that it increases student satisfaction and enjoyment, provides an example of abstract ideas presented in a lecture, allows for application of material and collaborative problem-solving, increases knowledge retention, and creates a common frame of reference and observational base that can be built upon throughout a course or degree program (Platt 1959; Harrison and Luithlen 1983; Walcott 1999; Wheeler 2001; Hefferan, Heywood, and Ritter 2002; Martin 2003; Abbott 2006). More specific learning outcomes that fieldwork can create are the development of observational skills, the facilitation of experiential learning, encouragement of students to take responsibility for their own learning, development of analytical skills, experience in real research, and development of respect for the environment (Gold et al. 1991). Thus, it is clear that field instruction can be a powerful tool that allows an instructor to increase the effectiveness of a course and successfully achieve many goals of educational quality.

For new instructors, there is a wealth of guidance available for developing field study (Hart 1968; Lounsbury and Aldrich 1986; Gold et al. 1991; Clark 1996; Kent, Gilbertson, and Hunt 1997; Livingston, Matthews, and Castley 1998; Bednarz 1999; Gerber and Chuan 2000; Rice and Bulman 2000; Healey and Roberts 2004). Perhaps one of the best references for initial creation of field studies is Kent, Gilbertson, and Hunt (1997), which outlines the different types of fieldwork and different approaches to fieldwork, and describes how to organize a field study through the broad steps of preparation, practice, and debriefing. Of equal value is the field study chapter in Gold et al. (1991), which provides highly detailed steps for the development of field studies.

As the diversity of student population in higher education increases, it is important to consider how field study can include and be welcoming to all students irrespective of their physical and mental disabilities; part-time and full-time enrollment; and difference in age, gender, nationality, sexuality, religion, family status, culture, language, and regional origins among many other markers of identity. For example, field study activities in different urban neighborhoods may be perceived as safe by some students, but dangerous to others. Some students, by reason of language difference, nationality, or gender, may not be able to undertake some qualitative survey methods. Part-time and working students or students with children can have difficulty participating in weekend or residential field study programs away from campus. Regarding students faced with limited mobility, visual impairment, hearing impairment, mental health difficulties, and hidden disabilities, Healey et al. (2001) offer an extensive guide. However, more attention to these issues is needed. Field study is, after all, one of the activities that often attracts students to geography for additional classes and as majors. Inclusive field experiences are a key way to make geography more welcoming to all students irrespective of their backgrounds and identities.

SPECIAL CHALLENGES OF FIELD STUDY AND HUMAN GEOGRAPHY

If field instruction is declining in geography programs, it has declined the most in human geography (Kent, Gilbertson, and Hunt 1997; Salter 2001). Time constraints and budget concerns are certainly factors in this decline, but the nature of fieldwork in human geography may also be a consideration. Human geography fieldwork depends on a great deal of overtly qualitative methodologies, such as observation and "thick" description, participant observation, ethnography, interviewing, and personal reflection. As researchers, we invest much effort in learning and sharpening our qualitative research skills that over time can become intuitive (Herbert 2000; Denzin and Lincoln 2000; Delyser 2001). As field instructors, however, it is easy to forget that we worked so hard and invested much time into learning those skills. Therefore, when designing field-based activities that depend heavily on qualitative data collection, the instructor must design a field exercise that is appropriate to the level of skill that students have, or can develop in a short period of time. A corollary to this is that the pre-field class time is critical because students have to be clearly instructed before going out to the field.

While the need for such detailed planning sounds obvious, the necessity is often learned through failure. To illustrate this point, consider the following example of a field exercise assigned to an undergraduate urban geography class:

> **Field Exercise: Riding down Main Street in Downtown Houston**
> Take a trip to downtown Houston and ride the seven-mile Main Street METRORail line. Make sure to notice how the landscape changes along the way. After your Metro journey is finished, pick a two- or three-block section of Main Street and walk around. How is the perspective different (on foot versus in the train)? What is happening there? Who is there with you? How is the space being used? Drawing from our discussion of cultural landscape, write about this urban cultural landscape.

The original goal of this assignment was to provide an urban experience for the students while also introducing key themes of the course. Not surprisingly, students felt this exercise to be frustrating. Their experiential observations were superficial and irrelevant to course content. They were unable to meaningfully compare different parts of the city. After more discussion with the students, it became clear that they were ill-prepared for a field-based experience. The content of the field assignment was there (although not as clearly stated as it could be), but the field "technique" aspect was missing. The students were sent out into the field without any tools or instructions on how to use those tools. Unfortunately, this is often the case with field exercises in human geography courses because as instructors we may assume that the necessary skills are somehow intuitive. When we reflected on this field assignment, we identified a number of ways to improve the experience for students, as follows.

Logistics

Detailed and clear directions about the logistics of field exercises are critical. In the above assignment, the students are sent out alone, without a field guide or instructor, with no specific directions about "how to" ride the METRORail such as where the station is, how much it costs, or where they can park. Particularly in an undergraduate setting, this lack of detail is not a good idea; as students were so distracted by trying to work out the details, they were unable to observe and experience the places on the trip. A good field exercise should be so well planned that when students are in the field, they are able to focus solely on the experience and work associated with being in the field.

Background Preparation

The Main Street exercise was assigned during the first few weeks of class before students had any contextual grounding in urban geography. The hope was that the trip would spark questions and ideas in their minds that would enrich class time. This strategy may have worked for graduate students but did not work for undergraduates. Therefore, it is essential that you consider the skills and experience of your students before implementing a field component.

Preparing students for the field takes significant classtime preparation (Gold et al. 1991). Readings, lectures, discussions, and even practice field trips around campus are vital for a successful field experience. Just as an instructor in a physical geography field exercise would demonstrate how to use field equipment, so must the human geographer clearly demonstrate his or her tools and techniques. Maps, old photographs, tourist brochures, cameras, sketchbooks, interviews, and surveys are just a few examples of the tools that could enhance such a field exercise. Students must be reminded, however, that their observational skills are the fundamental tools for any field experience, and using those skills must be intentional. Practicing field observation on campus (in the cafeteria, student union, or quadrangle) before the field exercise is an excellent strategy for making students more comfortable. A sample exercise of "Practicing Landscape Interpretation" on your campus is available on the web site for this book.

Content and Learning Objectives

The above assignment fails to give the students any key concepts, themes, or objectives around which to structure their field experience. Not only should you clearly discuss these with the students prior to the field trip, they should have a field sheet to take with them that also clearly states the objectives and themes of the exercise. Clearly identifiable learning objectives are vital to a successful field exercise (Gold et al. 1991; Jenkins 1994). However, a good field instructor will always leave room for the unexpected, even if that means being asked difficult or unanswerable questions. In fact, such an opportunity allows you to model good fieldwork—an extremely effective technique in teaching field methods. Uncertainty, while often uncomfortable, is rarely a detriment to learning and sometimes sets the stage for a broader range of learning outcomes like those discussed in the introduction to this chapter (Salter 2001; Fink 2003).

Assessment and Product

Finally, the above assignment has no clear product to present for a grade. Initially, the product was envisioned as a journal-style writing assignment that described the field experience in detail, including all of the wondrous insights had on the journey. Again, failure to adequately prepare the students resulted in weak final products. Having students keep a journal during a field course or exercise is a powerful tool for gauging student learning; however, most relevant literature makes it clear that students must be guided in their journal writing with specific discussion questions, ideas, and regular instructive feedback (Lewis and Mills 1995; Cantrell, Fusaro, and Dougerty 2000; Park 2003). For the instructor, it is important to remember your own struggles with journal writing and that it is a skill that takes time to develop.

For a short field exercise like this one, it would perhaps be better to assign a different project such as a photographic essay, or "mental map" exercise in which students create an annotated map of their day in the field. Whatever the assignment, students must be given clear instructions regarding what is expected and how it will be graded *before* the field experience. With this out of the way, students will relax and focus on being in the field. Again, a well-crafted field exercise will enable students to focus on observing and assessing what they are experiencing. (A revised portion of the Main Street field exercise is provided in the web site of this book.)

Teaching in the Field

Cary Komoto

Here are some suggestions on how the development of the urban geography field exercise could potentially be transformed into a scholarship of teaching and learning (SoTL) project. One way to explore the question of how effective the urban geography field exercise is in helping students meet a learning objective such as "to be able to interpret the cultural landscape of a city." Like traditional research, there should be a review of relevant literature to inform the research design. The research should include appropriate assessments to collect

(continued)

Continued

the necessary data for analysis. In addition, the researcher should determine what type of product will result from the project. Again, this may be a peer-reviewed publication, but in this case it might be a field trip guide and an associated commentary posted on a web site.

Recommended Readings

Fuller, I., S. Edmondson, D. France, D. Higgitt and I. Ratinen. 2006. International perspectives on the effectiveness of geography fieldwork for learning. *Journal of Geography in Higher Education* 30(1):89–101.

Haigh, M., and J. Gold. 1993. The problems with fieldwork: a group-based approach towards integrating fieldwork into the undergraduate geography curriculum. *Journal of Geography in Higher Education* 17(1):21–32.

Kent, M., D. Gilbertson, and C. Hunt. 1997. Fieldwork in geography teaching: A critical review of the literature and approaches. *Journal of Geography in Higher Education* 21(3):313–32.

Lonergan, N., and L. Andresen. 1988. Field-based education: Some theoretical considerations. *Higher Education Research & Development* 7(1):63–77.

SPECIAL CHALLENGES OF FIELD STUDY AND PHYSICAL GEOGRAPHY

Field instruction has long played a central role in physical geography education. Perhaps the main reason for this role is that it is impossible to completely represent the complexity and holistic nature of physical landscapes in the lecture hall. Consequently, many authors and studies have indicated that fieldwork is the best strategy for students to directly observe, understand, and appreciate physical processes (MacKenzie and White 1982; Kern and Carpenter 1986; Karabinos, Stoll, and Fox 1992; Malone 1999; Tueth and Wikle 2000; Hudak 2003; Abbott 2006). Further, fieldwork allows for representation of physical landscapes as a multidisciplinary concept (climatology and geomorphology) (Brown 1969) and assists students with setting complex physical interactions within a familiar local context (Eberhart and Thomas 1991; Fuller, Rawlinson, and Bevan 2000). The end result is that the field experience is very often characterized by both students and faculty as one of the most satisfactory learning experiences in their college career (Kent, Gilbertson, and Hunt 1997; Bodycott and Walker 2000; Drummond 2001; Stanitski and Fuellhart 2003).

As already outlined in this chapter, development of a successful field study activity or course relies upon extensive preparation that should focus on three components: pre-trip activities and orientation with students, in-the-field concerns, and post-trip activities and feedback (Kent and Gilbertson 1997). It is impossible to stroll into a small grove of trees with a box of instruments, tell the students to collect data, and expect a successful field activity. A successful field activity needs to be placed within the context and goals of a course before entering the field, and an opportunity must exist for students to reflect upon, discuss, or present the lessons learned from the activity. Thus, when developing the activity make sure to ask yourself how the activity fits into the grand scheme of the class and make sure students are aware of this big picture (Gold et al. 1991; Walcott 1999; Martin 2003). The remainder of this section will discuss issues that we have found to be particularly relevant to the development of physical geography field studies.

Exotic Landscape Syndrome

The first issue is what we call the "exotic landscape syndrome." Textbooks and similar resources are often accompanied by wonderful multimedia supplements that include pictures and graphics of specific phenomena. Usually these pictures and graphics are perfect examples of the physical landscape and can be quite useful in illustrating concepts (*e.g.*, the Grand Canyon as a fluvial landscape). However, an over reliance upon such perfect or exotic landscapes can create an under appreciation of the local landscape. Remember there are plenty of examples of the physical landscape on campus or in your community that represent course material just as effectively as publisher-provided graphics. Abbott (2006), Hudak (2003), and Jennings and Huber (2003) all offer excellent examples of how to utilize campus and local landscapes in the study of physical geography. An added benefit of such local field studies is that local governments or government officials can use collected data to develop plans and policies, creating an opportunity for students to apply classroom material to real-world problems (Jennings and Huber 2003).

Equipment and Materials

The second piece of advice addresses a common obstacle: purchasing equipment and materials for field studies. Introductory physical geography laboratories often require equipment such as topographic maps, compasses, sling psychrometers, soil testing kits, or sediment sieves; upper level and graduate physical geography classes require more advanced equipment such as Global Positioning System (GPS) units, surveying equipment, weather stations, and even sample analysis. Without adequate start-up funds or departmental resources to purchase such materials and equipment, many new instructors feel they have no alternative but to provide "canned" data for analysis or simply remove field components from their classes. The end result is that students take a physical geography class without ever being exposed to the challenge of collecting quality primary data, a skill required for a deep understanding of physical geography.

When faced with such a challenge, you should remember that many avenues exist for circumnavigating the lack of funding for field materials and equipment. First off, do not overlook the power of simple observation. Meaningful landscape analysis can be completed with sketch maps, pacing off distances, using cell phone cameras, and estimates of height. Secondly, many cheap and free options exist for acquiring field materials and equipment. Challenge your students to search the Internet for free digital copies of maps or air photos. We have found that students are resourceful at finding materials, and through such a task students feel truly engaged in the course. In terms of equipment, send a campus-wide e-mail asking other faculty if they have any old field equipment they no longer need. You will be surprised how many people are willing to get junk out of their office. Finally, equipment does not always have to be bought. Sometimes it can be built relatively quickly and cheaply (Pease et al. 2002a, 2002b). Involving students in the construction of field equipment is another avenue to engaging students in a course.

Pre- and Post-Field Work

The last piece of advice for developing field study in physical geography is this: do *not* forget the time before and after venturing into the field. We love being in the field and jump at any chance to get out of the office. We have also found that the majority of students enjoy the field more than the classroom. However, do not allow this desire to rush into the field diminish preparation or analysis/reflection associated with field study. As a general rule of thumb, we schedule at least the same amount of time for preparation and analysis/reflection as we do for actual field instruction. The field instruction time does not include travel time such as hiking or driving, but only the time students are engaged in an activity in the field. Without scheduling the pre- and post-field study periods, we have found, and existing literature supports this belief, students are unable to truly grasp the links between field analysis and course content and are just "going through the motions" of collecting data in the field (Harrison and Luithen 1983; Lonergan and Andresen 1988; Haigh and Gold 1993; Kent, Gilbertson, and Hunt 1997; Jennings and Huber 2003).

For example, for a ten-day field trip in the Bahamas (conducted over winter break or the first summer term), trial and error was needed to find that the most effective format for the pre-field study component was an on-campus class in Caribbean geography, history, or political science. Such a course provides a background and context within which students can place their field observations and is a requirement for the field class. Then once in the field in the Bahamas, the daily routine is field study from 8 A.M. to 3 P.M., then a discussion of the day's activities until 5 P.M., followed by dinner and data analysis from 7 P.M. to 9 P.M. It makes for a long day, but students are still given time for relaxation and recreation. After the field component in the Bahamas, students are required to hand in a field report or research paper at the end of the term. This allows for a three- to six-week

Virtual Field Studies

Cathryn Springer

Virtual Field Studies (VFS) are technologically mediated experiences of places or field sites (Stainfield et al. 2000; Springer 2007). They vary in levels of user-control and interactivity, degree of realism, type of investigation, and amount and type of sensory stimulation. Geographers and other educators have proposed multiple learning objectives for VFS, and VFS research has pointed to a number of cognitive and affective gains among learners (Warburton and Higgitt 1997; Crampton 1999; Foskett 2000; Spicer and Stratford 2001; Springer 2007; Stainfield et al. 2000).

Geography educators generally use VFS to augment rather than replace traditional field studies, but the rising pressures of liability, cost, and time associated with conducting field studies at an actual field location have caused many to view VFS as alternative options for incorporating field experiences into the geography curriculum (Haigh and Gold 1993; McEwen 1996; McEwen and Harris 1996; Kent, Gilbertson, and Hunt 1997; Holmes and Thomas 2000; Stainfield et al. 2000).

What Are Some Effective Practices with VFS?

Research suggests that the learning quality of VFS is linked to several key characteristics, which include the quality and nature of interaction, visual appeal, immediate learning environment, structure and flow of the lesson, and personalization of material (Springer 2007). Specifically, the following practices are important for maximizing the effectiveness of VFS:

- Incorporate meaningful mental and physical interactivity
- Explicitly state the purpose(s) of the lesson
- Personalize the lesson to students' experiences and interests, which can be done through inquiry-based education
- Ensure student control of learning
- Allow sufficient time for varied paces of students and for individual exploration
- Have a clear and simple structure to the lesson
- Check the web site or activity for easy navigation, and working and useful links (if applicable)
- Create a simple handout, if necessary, that shows what features are available in the VFS, for example, a measuring tool, or a way of adjusting resolutions of images
- Choose VFS that produce opportunities for students to be challenged by new information
- Use VFS with high visual appeal in digital representations and/or web sites
- Have some sort of "field guide," definite task to accomplish, or set of steps to aid students in activities and/or interpretations—these can also provide immediate feedback to students for building confidence and challenging any misconceptions they may have

(continued)

Continued

What Are Some Good VFS Resources?

The web offers large amounts of resources for VFS, which can make it difficult to know where to begin. Three good (and free!) resources readily available are NASA World Wind, Google Earth, and the Degree Confluence Project. Both NASA World Wind (http://worldwind.arc.nasa.gov/) and Google Earth (http://earth.google.com/) consist of satellite imagery for the entire world. NASA World Wind offers different satellite overlays, including a three-dimensional (3-D) image to be used with 3-D glasses. Google Earth, meanwhile, features a measuring tool, additional overlays for download, various layers to make active (including 3-D), and a toggle/tilt feature that allows for a side view of a landscape in addition to a bird's eye view. The degree confluence project at www.confluence.org is a well-established web site of images and journal entries from people who have visited intersections of major degrees of latitude and longitude. This web site stands out from others because the journal entries and snapshots provide personal stories and information about the location, while providing users with control for navigating to different locations. Possible student activities with these and other VFS resources include the following:

- Measuring distances, sizes, and areas, such as of floodplains and commuting distances
- Counting occurrences of particular phenomena (i.e., number of oxbow lakes in a particular region, or the number of certain cultural features)
- Sketching the landscape, particular phenomena, and maps
- Graphing elevations, densities, slope
- Classifying phenomena, such as type of channel patterns, landforms, economic sectors in a city, settlement patterns
- Identifying important features in the human and physical landscape
- Comparing satellite imagery with topographic maps of the same area

period when the students work with me (Doug) to complete their analysis. Usually, the end result is a high-quality research project that offers insight based on rigorous data analysis. Obviously, such a format does not work for short-term field activities, but the structure of preparation, field study, and analysis/reflection can be successfully adapted to any temporal scale of field instruction.

CONCLUSION

Teaching in the field is a rich and rewarding experience for both students and instructors that can address multiple learning objectives. As can be seen from this discussion, field instruction in human and physical geography can be different in terms of targeted learning outcomes and the use of equipment. However, there are common themes that are salient to all geographic field instruction. These issues include thorough pre-fieldtrip preparation; the linking of field instruction with clear learning objectives and themes; demonstration of tools and techniques; the overlooked utility of local field trips; and the need for a post-fieldtrip debriefing that clearly links activities to course objectives, assessment, and expected outcomes.

In short, the responsibility of the field instructor is to take care of all logistical details and clearly define objectives so that the student can focus solely on observation in the field. Even with great preparation, however, field experiences will sometimes not go as well as planned. With this, the advice of veteran professors rings true. Successful fieldwork is sometimes a trial-and-error process that can be affected by forces beyond our control (unbearable heat, a flat tire, a museum suddenly closed, or awkward group dynamics). The challenge is to maintain an enthusiasm for and commitment to teaching in the field even though it requires much effort and time. When it works, the rewards are great—for both the student and teacher.

References

Abbott, J. A. 2006. Measuring thermal variations in a valley using a team, field project designed by students. *Journal of Geography* 105(3):121–28.

Bednarz, S. W. 1999. Fieldwork in K-12 geography in the United States. *International Research in Geographical and Environmental Education* 8(2):164–70.

Bodycott, P., and A. Walker. 2000. Teaching abroad: Lessons learned about intercultural understanding for teachers in higher education. *Teaching in Higher Education* 5(1):79–94.

Brown, E. H. 1969. The teaching of fieldwork and the integration of physical geography. In *Trends in Geography*, eds. R. U. Cooke and J. H Johnson, 70–78. Oxford: Pergamon.

Buckley, G. L., N. R. Bain, A. M. Luginbuhl, and M. L. Dyer. 2004. Adding an "active learning" component to a large lecture course. *Journal of Geography* 103:231–37.

Cantrell, R. J., J. A. Fusaro, and E. A. Dougerty. 2000. Exploring the effectiveness of journal writing on learning social studies. *Reading Psychology* 21(1):1–11.

Clark, D. 1996. The changing national context of fieldwork in geography. *Journal of Geography in Higher Education* 20(3):385–91.

Crampton, J. W. 1999. Integrating the web and the geography curriculum: The Bosnian virtual fieldtrip. *Journal of Geography* 98:155–68.

DeLyser, D. 2001. "Do you really live there?" Thoughts on insider research. *Geographical Review* 91(1–2):441–53.

DeLyser, D., and P. F. Starrs. 2001. Doing fieldwork: Editor's introduction. *Geographical Review* 91:4–9.

Denzin, N. K., and Y. S. Lincoln, eds. 2000. *The handbook of qualitative research.* Thousand Oaks, CA: Sage.

Drummond, D. 2001. Travel abroad: A *sine qua non* for geography teachers. *Journal of Geography* 100:174–75.

Eberhart, L. L., and J. M. Thomas. 1991. Designing environmental field studies. *Ecological Monographs* 61:53–73.

Fink, D. 2003. *Creating significant learning experiences: An integrated approach to designing college courses*. San Francisco: Jossey-Bass.

Foskett, N. 2000. Fieldwork and the development of thinking skills. *Teaching Geography* 25(3):126–29.

Fuller, I., S. Rawlinson, and R. Bevan. 2000. Evaluation of student learning experiences in physical geography fieldwork: paddling or pedagogy? *Journal of Geography in Higher Education* 24(2):199–215.

Gerber, R., and G. K. Chuan, eds. 2000. *Fieldwork in geography: Reflections, perspectives, and actions*. Dordrecht, The Netherlands: Kluwer Academic Publishers.

Gold, J. R., A. Jenkins, R. Lee, J. Monk, J. Riley, I. D. H. Shepherd, and D. J. Unwin. 1991. *Teaching geography in higher education: A manual of good practice*. Oxford, U.K.: Blackwell.

Haigh, M., and J. R. Gold. 1993. The problems with fieldwork: A group-based approach towards integrating fieldwork into the undergraduate geography curriculum. *Journal of Geography in Higher Education* 17(1):21–32.

Harrison, C., and L. Luithlen. 1983. Fieldwork for land use students: An appraisal. *Journal of Geography in Higher Education* 7(1):23–32.

Hart, J. F. 1968. The undergraduate field course. In *Field training in geography*, ed. J. F. Hart, 29–38. Washington, DC: Association of American Geographers, Commission on College Geography, Technical Paper No. 1.

Healey, M., A. Jenkins, J. Leach, and C. Roberts. 2001. *Issues in providing learning support for disabled students undertaking fieldwork and related activities*. Cheltenham, U.K.: University of Gloucestershire, Geography Discipline Network.

Healey, M., and Roberts, J., eds. 2004. *Engaging students in active learning: Case studies in geography, environment, and related disciplines*. Cheltenham, U.K.: University of Gloucestershire, Geography Discipline Network.

Hefferan, K. P., N. C. Heywood, and M. E. Ritter. 2002. Integrating field trips and classroom learning into a capstone and undergraduate research experience. *Journal of Geography* 101:182–90.

Herbert, S. 2000. For ethnography. *Progress in Human Geography* 24(4):550–68.

Higgitt, M. 1996. Addressing the new agenda for fieldwork in higher education. *Journal of Geography in Higher Education* 20(3):391–98.

Holmes, D., and T. Thomas. 2000. Fieldwork and risk management. *Teaching Geography* 25(2):71–74.

Hudak, P. F. 2003. Campus field exercises for introductory geoscience courses. *Journal of Geography* 102:220–25.

Jenkins, A. 1994. Thirteen ways of doing fieldwork with more students. *Journal of Geography in Higher Education* 18(2):143–54.

Jennings, S. A., and T. P. Huber. 2003. Campus-based geographic learning: A field oriented teaching scenario. *Journal of Geography* 102:185–92.

Karabinos, P., H. M. Stoll, and W. T. Fox. 1992. Attracting students to science through field exercises in introductory geology courses. *Journal of Geological Education* 40:302–5.

Kent, M., D. Gilbertson, and C. Hunt. 1997. Fieldwork in geography teaching: A critical review of the literature and approaches. *Journal of Geography in Higher Education* 21(3):313–32.

Kern, E. L., and J. R. Carpenter. 1986. Effect of field activities on student learning. *Journal of Geological Education* 34:180–83.

Lewis, S., and C. Mills. 1995. Field notebooks: A student's guide. *Journal of Geography in Higher Education* 19(1):111–14.

Livingstone, I., H. Matthews, and A. Castley. 1998. *Fieldwork and dissertations in geography*. Cheltenham, U.K.: University of Gloucestershire, Geography Discipline Network.

Lonergan, N., and L. W. Andresen. 1988. Field-based education: Some theoretical considerations. *Higher Education Research and Development* 7(1):63–77.

Lounsbury, J. F., and F. T. Aldrich. 1986. *Introduction to geographic field methods and techniques*, 2nd ed. Columbus, OH: Charles E. Merrill Publishing Company.

MacKenzie, A. A., and R. T. White. 1982. Fieldwork in geography and long term memory structures. *American Educational Research Journal* 19(4):623–32.

Malone, D. H. 1999. A faculty survey on field trips in undergraduate structural-geology courses. *Journal of Geological Education* 47:8–11.

Martin, D. G. 2003. Observing metropolitan Atlanta, Georgia: Using an urban field study to enhance student experiences and instructor knowledge in urban geography. *Journal of Geography* 102:35–41.

McEwen, L. 1996. Fieldwork in the undergraduate geography programme: Challenges and changes. *Journal of Geography in Higher Education* 20(3):379–84.

McEwen, L., and F. Harris. 1996. The undergraduate geography fieldweek: Challenges and changes. *Journal of Geography in Higher Education* 20(3):411–21.

Nairn, K., D. Higgitt, and D. Vanneste. 2000. International perspectives on field-courses. *Journal of Geography in Higher Education* 24(2):246–54.

Park, C. 2003. Engaging students in the learning process: The learning journal. *Journal of Geography in Higher Education* 27(2):183–99.

Pease, P., S. Lecce, P. Gares, and M. Lange. 2002a. Suggestions for low-cost equipment for physical geography I: Laboratory equipment. *Journal of Geography* 101(4):167–75.

———. 2002b. Suggestions for low-cost equipment for physical geography II: Field equipment. *Journal of Geography* 101(5):199–206.

Platt, R. S. 1959. *Field study in American geography: The development of theory and method exemplified by selections*. Chicago: University of Chicago, Department of Geography, research paper no. 61.

Rice, G. H., and T. L. Bulman. 2000. *Fieldwork in the geography curriculum: Filling the rhetoric-reality gap*. Jacksonville, AL: National Council for Geographic Education.

Rundstrom, R. A., and M. S. Kenzer. 1989. The decline of fieldwork in human geography. *Professional Geographer* 41(3):294–303.

Salter, C. L. 2001. No bad landscape. *The Geographical Review* 91(1–2):105–12.

Sauer, C. O. 1956. The education of a geographer. *Annals of the Association of American Geographers* 46:287–99.

Spicer, J. I., and J. Stratford. 2001. Student perceptions of a virtual field trip to replace a real field trip. *Journal of Computer Assisted Learning* 17:345–54.

Springer, C. E. 2007. The construction of place in technologically-mediated learning environments: The nature, process, and effects of virtual field studies. Unpublished Ph.D. dissertation, Texas State University—San Marcos, San Marcos, Texas.

Stainfield, J., P. Fisher, B. Ford, and M. Solem. 2000. International virtual field trips: A new direction? *Journal of Geography in Higher Education* 24(2):255–62.

Stanitski, D., and K. Fuellhart. 2003. Tools for developing short-term study abroad classes for geography studies. *Journal of Geography* 102:202–15.

Tueth, M. W., and T. A. Wikle. 2000. The utility and organization of a college field course: Examining national park management. *Journal of Geography* 99:57–66.

Walcott, S. M. 1999. Fieldwork in an urban setting: Structuring a human geography learning experience. *Journal of Geography* 98:221–28.

Warburton, J., and M. Higgitt. 1997. Improving the preparation for fieldwork with "IT": Two examples from physical geography. *Journal of Geography in Higher Education* 21(3):333–47.

Wheeler, J. O. 2001. Traditional versus contemporary fieldwork in American urban geography. *Urban Geography* 22:95–99.

CHAPTER 12

Geography and Global Learning

Michael Solem

"Think globally, act locally"—a slogan coined by French microbiologist René Dubos—is now a call to action to conserve the earth's environment. But what does it mean to *think globally*? Is it merely a catchphrase or does it have deeper meaning? Is there really a method of thinking globally, one that can be developed via formal instruction much like critical or creative thinking?

If the nineteenth-century writings of Russian geographer Piotr Kropotkin are any clue, then it seems safe to say that geographers have been reflecting on global thinking for quite some time. Teaching as he was in an era when geographic research was often applied and cited in support of Western imperialism, Kropotkin's writings attempted to steer the discipline in an opposite course. Geography, he implored, should "teach that all men [sic] are brethren, whatever be their nationality" (1885: 943). I think Kropotkin would agree that achieving this ideal requires more than a course in world regional geography—it implies rethinking the way we infuse global perspectives throughout the curriculum.

Recent calls to address this need—to "internationalize" college campuses and curricula—is a trend that many scholars expect to become a defining element of twenty-first-century higher education (Haigh 2002). But a problem is that internationalization carries many meanings that are used inconsistently from college to college and from discipline to discipline. Sometimes the term refers to applied practices such as joint-degree programs between domestic and foreign universities. Other institutions use the term to describe their student exchange programs or study abroad courses. In still other cases, internationalization is considered to be the process of developing curricula for promoting global perspectives in liberal education (Hovland 2006). As can be seen from the additional resources listed at the end of this chapter, geography is rich in activities like these that enhance international collaboration and widen opportunities for students to become more aware and knowledgeable of global dynamics and issues (Reeve et al. 2000; Haigh 2002). But can we, as geography teachers, do more in our classrooms? This call for internationalization seems to imply that we also rethink the way we develop learning activities for our students and the way we approach our teaching.

In this chapter, my principal aim is to demonstrate how you can help your students learn to think globally by developing your ability to *teach* globally—that is, by drawing on the expertise, resources, and collegiality of the international community of geographers in higher education. But first, let us consider some of the foundational principles of international education and their relevance for teaching geography in higher education. Armed with that knowledge, you can get started with developing your own global learning activities by completing Activity 12.1 on this book's web site. This activity will also introduce you to some international learning networks in geography, which I've found to be valuable for building educational collaborations with international colleagues.

BEING AND BECOMING A GLOBAL CITIZEN

Perhaps the fundamental purpose of international education is to cultivate in learners a sense of responsibility for the world and its varied cultures and environments—in other words, to instill an ethic of global citizenship (Hayden and Thompson 1995). Global citizenship begins with the knowledge that the actions of individuals and nations can have global consequences and encompasses the duties, rights, and privileges that stem from this knowledge. Even seemingly innocuous activities such as purchasing sneakers or coffee at a local store can have a direct impact on the lives of workers living thousands of miles away. Understanding that every person, through a global web of economic, environmental, and cultural strands, is connected to people and places near and far is an essential perspective for life and work in a rapidly globalizing and interdependent world. It is, simply put, an example of thinking globally.

Since at least the mid-1990s it has become widely accepted in the scientific community that national policy responses alone are not sufficient to solve world geographical problems (Rediscovering Geography Committee 1997). International cooperation is now necessary for achieving long-term, sustainable, and workable solutions. Similarly, teaching geography can benefit from approaches that expand the borders of traditional learning spaces by linking together students from different countries for international learning and problem solving activities. By learning in this way, students can contribute to the geographic knowledge and intercultural experiences of their peers. Yet many students lack these experiences as part of their geography education, for several reasons. First, less than 25 percent of geography professors in the U.S. have ever been involved with an educational project involving some form of international collaboration or interaction (Ray and Solem forthcoming). Second, less than 3 percent of American undergraduates participate in study abroad programs, and those who do so mostly travel overseas for less than a month in a Western European country such as the United Kingdom or France with relatively similar cultural histories, languages, or economic systems (Green and Olson 2003). Still other students attend campuses with low populations of international students or are unable to overcome financial obstacles for overseas study.

The present situation in geography is therefore quite ironic: Although much of the content of the undergraduate geography curriculum is demonstrably international in focus, the actual *process* of teaching that content is largely performed by the individual professor and rooted in the perspective of that person's nationality. Given this context, what can be done to provide more geography students with opportunities to gain global perspectives and international learning experiences during their college education? In a globalizing world, what are our obligations as geography educators for preparing students to think globally and develop intercultural skills that will later help them succeed in an increasingly diverse and global workplace? Is it sufficient to merely introduce those students to geographical facts, or does becoming a global citizen require students to go one step further and learn to live and cooperate with people from a different culture?

Regardless of whether we choose to address explicitly the values dimension of geography in our teaching, believing that your students would benefit from having a global perspective about any issue *is* a value judgment, just as would be a decision to ignore the views of other cultures in favor of an unwavering focus on the national perspective (Reeve et al. 2000; Alexander 2001). By revealing for your students the underlying similarities and differences of their personal experiences in relation to peoples and places elsewhere, you will at the very least contribute to their intellectual understanding of how the outcomes of geographic processes can vary from place to place and from region to region. Although it may be difficult to know whether your students ultimately develop a greater appreciation for different cultures or a sense of global responsibility as a result of gaining this knowledge, it seems that a starting point for educating future global citizens would require nothing less.

What, then, should a global citizen know, be able to do, and care about? There is neither a consensus among educators, nor is there room in this chapter to provide a thorough overview of what others have written on the issue. As an entry point into this literature, I would simply like you to consider the ideas of Robert Hanvey, a scholar who authored a widely cited 1976 essay which argued that any student can learn to view the world and its inhabitants from a global perspective—that is, a worldview marked by an understanding of and sensitivity toward world cultures and environments. Take a minute to

FIGURE 12.1 Five dimensions of a global perspective (Hanvey 1976)

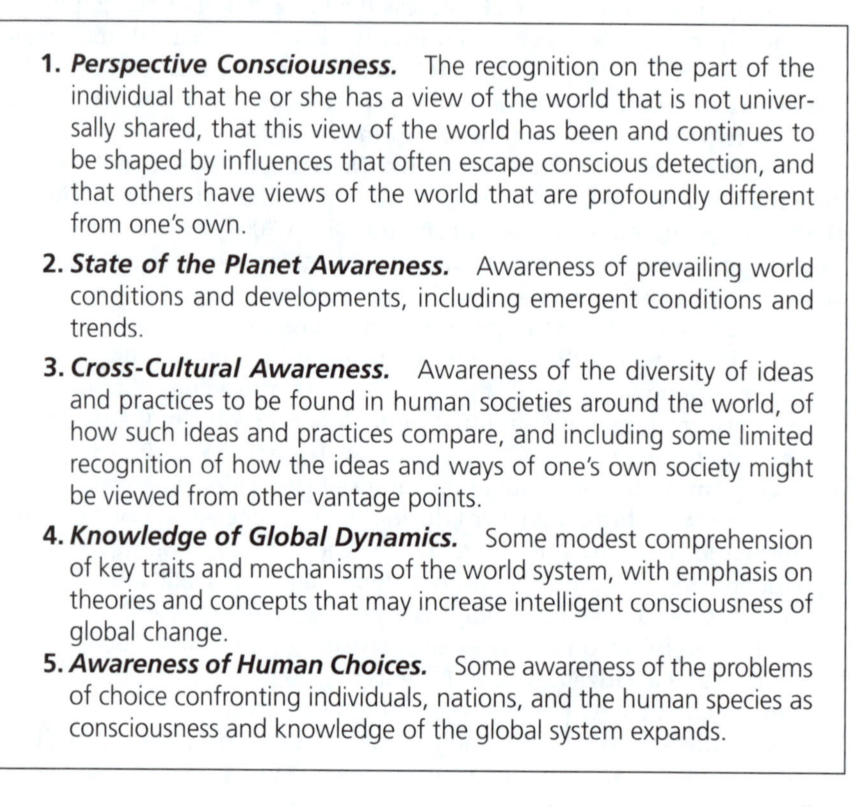

1. ***Perspective Consciousness.*** The recognition on the part of the individual that he or she has a view of the world that is not universally shared, that this view of the world has been and continues to be shaped by influences that often escape conscious detection, and that others have views of the world that are profoundly different from one's own.
2. ***State of the Planet Awareness.*** Awareness of prevailing world conditions and developments, including emergent conditions and trends.
3. ***Cross-Cultural Awareness.*** Awareness of the diversity of ideas and practices to be found in human societies around the world, of how such ideas and practices compare, and including some limited recognition of how the ideas and ways of one's own society might be viewed from other vantage points.
4. ***Knowledge of Global Dynamics.*** Some modest comprehension of key traits and mechanisms of the world system, with emphasis on theories and concepts that may increase intelligent consciousness of global change.
5. ***Awareness of Human Choices.*** Some awareness of the problems of choice confronting individuals, nations, and the human species as consciousness and knowledge of the global system expands.

examine Hanvey's ideas in Figure 12.1—can you think of ways that geography can help students acquire the knowledge, skills, and traits described?

To stimulate your thinking further, consider the role that your selection of teaching methods might play in determining whether students acquire a global perspective in the geography classroom. Recall the point I made a little earlier about the process of teaching. This is an important issue, because internationalization is not only concerned with building intercultural content into undergraduate courses, but it is also aimed at supplying undergraduates with opportunities to interact personally with peers and experts from different countries. Traditionally, this interaction has been achieved through study abroad programs. But there are also other means of supporting international connections in the geography curriculum that can be achieved locally and via applications of media and technology.

One relatively straightforward technique would be to assign material written by authors from other countries to complement the textbooks and instructional materials by U.S. scholars the students typically read. These materials can include everything from novels and films by foreign authors to online newspapers, blogs, and other sources of information originating from outside the U.S. Aspaas (1998), for instance, proposed getting students involved in this process by having them work in small groups to seek out information on current events from international sources, comparing what they find to the content and perspectives of writings by authors and media outlets in the U.S. Editorial pages from world newspapers can provide interesting examples of public opinion on issues ranging from terrorism and gas prices to free trade and global warming.

Another way to get students involved in the internationalization process was suggested to me by Bradley Rink, a Ph.D. candidate in geography at the University of Cape Town and an international educator working as a resident director of a study abroad program in South Africa. Bradley's suggestion is to

> "utilize the experience of students abroad as a source of information." International educators (particularly those based overseas) are keenly aware of the level of connectivity that students abroad have with the home campus. Why not utilize those connections by engaging students on the home campus and those who are currently studying abroad? In a geography course on "Cities of the South," for example, it could be interesting to get data (including photographs) and impressions from students studying in Africa, South America, or Asia. In this way, the connections (e-mail, social networking web sites, etc.) that are already in place between students can serve to support an academic dialogue in the classroom and further add to global perspectives" (personal communication with Bradley Rink on 15 August 2007).

For additional examples, Fortuijn (2002) provides a good overview of issues arising in multicultural and international classrooms, and how that diversity can be tapped for maximum learning dividends.

Modern technology also makes it possible for geography faculty and students to communicate and collaborate across great distances (Hurley, Proctor, and Ford 1999; Kankaanrinta and Masalin 2001; Reed and Mitchell 2001), a "learning anytime, anywhere" approach that has prompted many educators to view the Internet as a powerful tool for internationalization (Solem et al. 2003). In geography, an example of this practice is the AAG Center for Global Geography Education (CGGE). With support from the National Science Foundation, the CGGE offers a collection of online course modules (*Population*, *Global Economy*, *Nationalism*, *Water Resources*, *Migration*, and *Global Climate Change*) supporting international collaboration in geography classrooms. The CGGE acts as a platform to promote international dialogue and learning partnerships for the study of world geographical problems. The idea is that undergraduates in any country can effectively learn about what is occurring there, and how people in those places react to those issues, by working collaboratively.

The CGGE modules, which are the focus of Activity 12.1, are intended for introductory and advanced undergraduate courses in geography. Each features several activities that encourage collaborative inquiry into geographic issues, placing emphasis on data sharing and discussion of issues by international teams of undergraduates. The modules are available in multiple languages and feature language translation tools to support bilingual collaboration. Asking U.S. undergraduates to practice communicating in a foreign language can be a good strategy for building their appreciation of global cultures and the challenges—and rewards—of international collaboration. It also aims to avoid "privileging" one language over another in the presentation and implementation of an international collaboration project (Samovar and Porter 2003). Note that this idea can be extended to any type of writing exercise by, for instance, asking students working on a geography assignment to write two or three paragraphs in their "second best language" (Daniels 2002).

Undergraduate students who have used the CGGE modules note the value of communicating directly about issues with their peers in other countries to learn about other perspectives. Group discussion of issues also requires students to reflect on their *own* perspectives in order to communicate them to people elsewhere (Friedsohn and Rubin 2002; Geores and Cirrincione 2002). Consider the words of a geography professor who participated in a CGGE project supporting undergraduate learning collaborations in the U.S. and Australia:

> For a subject such as geography where the focus is on global environments and human interaction with those environments, it is essential to examine international perceptions and values. For our students to hear opinions and exchange ideas and knowledge with students from another university allows our students to put into context the work we study in the classroom or lecture theatre. It shows them that the problems we face as an individual country are not that different from other countries and we all face similar challenges. (Klein 2005: 57)

But achieving these outcomes through technology can be tricky, and require a lot of preparation time on the part of the instructor and student to ensure that robust learning occurs from the international exchange. When the CGGE modules were tested in ten different countries between 2004 and 2005, the researchers discovered a number of factors that made it difficult for many students to enter into sustained, meaningful discussion of geographic issues resulting in greater awareness and appreciation of the perspectives held by their international teammates. In the CGGE project we tried to build upon these findings to improve the effectiveness of the modules, which you can review along with some of the evaluation results as you work on Activity 12.1.

Another important strategy for internationalizing your teaching and other areas of professional practice is to establish professional relationships

Evaluating the AAG Center for Global Geography Education

Cary Komoto

The Center for Global Geography Education (CGGE) modules provide a good illustration of how a mixed-method evaluation was designed and used to assess the effectiveness of a geography education project. In this instance, the CGGE modules were evaluated using a variety of methods in ten different countries. Specifically, over 500 students took part in the classroom trials, which varied in size and composition, during the 2004–2005 academic year. For example,

> *trials in Germany and China involved advanced students training to become geography teachers at pedagogical universities. Students in Chile and Northern Ireland were mainly geography majors with extensive coursework in the subject. Class sizes at these sites were small. By contrast, participants in Italy, Australia, and Spain were advanced students, but mostly nongeography majors, and most of the six U.S. trials were in larger, freshman-level introductory classes. U.S. students had less background in geography than their international counterparts.*
>
> *Data were collected from four main sources during the trials. First, students were twice administered (as a pre-test and post-test) two separate evaluation instruments. One instrument consisted of a set of eight short-answer content questions and one open-ended question seeking student reactions to international collaboration. The second instrument sought a variety of demographic information and asked students to answer a series of Likert questions about their attitudes toward studying these types of materials. The second source of data was a ten-item questionnaire e-mailed to the participating faculty, asking for their reactions to this experience. Third, for selected paired trials, the frequency and content of the international discussion board exchanges was analyzed. Fourth, the project evaluator made two site visits in Europe to observe and interview students and professors as they used the CGGE module with teammates in the U.S. (Klein and Solem forthcoming).*

This approach yielded a rich, albeit varied, set of data (see Activity 12.1). The analysis of the data was then used to inform the revision of the CGGE modules, which are available on the AAG web site at www.globalgeography.aag.org. CGGE is an illustration of SoTL because it started with a question about student learning that led to data collection, analysis, and ultimately publication and dissemination of tested instructional materials.

Recommended Readings

Klein, P., and M. Solem. forthcoming. Evaluating the impact of international collaboration on geography learning. *Journal of Geography in Higher Education*.

Palloff, R., and K. Pratt. 2005. Collaborating online: Learning together in community. San Francisco: Jossey-Bass.

Shepherd, I., J. Monk, J., and J. Fortuijn, J. 2000. Internationalization of geography in higher education: Towards a conceptual framework. Journal of Geography in Higher Education 24(2):285–98.

with geographers from around the world who share your teaching and research interests. In this way, as the American Council on Education notes, academic disciplines can take a leadership role in accelerating the pace of internationalization on college campuses (Green and Shoenberg 2006). Professional societies such as the AAG can provide an especially far-reaching and broad-based platform for acquiring and communicating information of significance to college and university faculty and instructors, providing them with educational resources, professional networks, and research opportunities for internationalizing their teaching and research practices. Students can also play an important role in the internationalization process by participating in educational programs such as study abroad and the international baccalaureate, as well as by being involved in informal education programs of the sort typically offered by international organizations on campus.

Two organizations, in particular, exist with the mission of building bridges that connect geographers around the world: the International Network for Learning and Teaching Geography in Higher Education (INLT) and the International Geographical Union (IGU). Some of the richest and most rewarding experiences in my professional career have been with INLT and IGU associates, many of whom have been a steadfast source of mentoring, encouragement, and intellectual exchange as I began my career. From these relationships I have developed a deep appreciation for geography and the way it is practiced in a wide variety of educational settings around the world. This has helped me to act locally to refresh my teaching with international methods and materials. I hope these ideas and resources will convince you to do the same and, in turn, set your students on a path toward global citizenship.

Additional Resources

International Network for Learning and Teaching Geography in Higher Education (INLT). INLT is a collaborative network of hundreds of geographers who share interests in research on teaching and learning in higher education. The INLT was established at the International Symposium for Learning and Teaching Geography in Higher Education held at the 1999 AAG Annual Meeting in Honolulu, with the aim of improving the quality and status of learning and teaching of geography in higher education internationally. INLT focuses on three areas of activity to:

1. promote innovative, creative, and collaborative research as well as critical reflection on learning and teaching of geography;
2. facilitate the exchange of materials, ideas, and experiences about learning and teaching of geography and to stimulate international dialogue;
3. create opportunities for geographers to network and form an inclusive, international community in higher education through meetings and workshops at major professional conferences, including the AAG, International Geographical Union (IGU), and the National Council for Geographic Education.

To receive the *INLT Newsletter* and keep in touch with discussions and news, subscribe to the INLT listserv by filling out the form available at: http://www.geog.canterbury.ac.nz/inlt/

International Geographical Union (IGU). IGU promotes geography research and education globally through publications, international conferences, and research projects. Established in Brussels in 1922, IGU features more than thirty-five commissions that provide opportunities for geographers to engage with colleagues sharing research interests in all major subfields of geography. Visit the IGU web site, http://www.igu-net.org, to learn about upcoming meetings and how to get involved in a commission related to your research specialty.

My Community, Our Earth: Geographic Learning for Sustainable Development. The My Community, Our Earth (MyCOE) project is a partnership to encourage youth to use geographic tools and concepts to address local issues of sustainability. By participating in MyCOE, young people from around the world can examine and learn more about environmental issues, determine patterns and trends, and propose solutions to the challenges they study.

You can get involved with MyCOE in several ways. One option is to become a MyCOE mentor. You will define the extent of help you wish to offer, but through e-mail, telephone, web chats, face-to-face visits, and/or other means of collaboration, you can make a difference in your geographic and specialty area. For additional opportunities, please visit the MyCOE homepage: www.geography.org/sustainable

References

Alexander, R. 2001. Border crossings: Towards a comparative pedagogy. *Comparative Education* 37(4):507–23.

Aspaas, H. 1998. Integrating world-views and the news media into a regional geography course. *Journal of Geography in Higher Education* 22(2):211–27.

Daniels, J. 2002. Writing across borders: An exercise for internationalizing the women's studies classroom. In *Encompassing gender: Integrating international studies and women's studies*, eds. M. M. Lay, J. Monk, and D. R. Rosenfelt, 404–12. New York: The Feminist Press.

Fortuijn, J. 2002. Internationalizing learning and teaching: A European experience. *Journal of Geography in Higher Education* 26(3):263–73.

Friedsohn, D., and B. Rubin. 2002. Talking Turkey: Cultural exchanges via e-mail between students at New Jersey City University and Bilkent University, Ankara. In *Encompassing gender: Integrating international studies and women's studies*, eds. M. M. Lay, J. Monk, and D. R. Rosenfelt, 322–34. New York: The Feminist Press.

Geores, M., and J. Cirrincione. 2002. Simulating cultural dissonance: A role playing exercise. In *Encompassing gender: Integrating international studies and women's studies*, eds. M. M. Lay, J. Monk, and D. R. Rosenfelt, 413–19. New York: The Feminist Press.

Green, M., and C. Olson. 2003. *Internationalizing the campus: A user's guide*. Washington, DC: American Council on Education.

Green, M., and R. Shoenberg. 2006. *Where faculty live: Internationalizing the disciplines*. Washington, DC: American Council on Education.

Haigh, M. 2002. Internationalization of the curriculum: Designing inclusive education for a small world. *Journal of Geography in Higher Education* 26(1):49–66.

Hanvey, R. 1976. An attainable global perspective. *Theory into Practice* 21(3):162–67.

Hayden, M., and J. Thompson. 1995. International schools and international education: A relationship reviewed. *Oxford Review of Education* 2(13):327–45.

Hovland, K. 2006. Shared futures: Global learning and liberal education. Washington, DC: Association of American Colleges and Universities.

Hurley, J., J. Proctor, and R. Ford. 1999. Collaborative inquiry at a distance: Using the Internet in geography education. *Journal of Geography* 98(3):128–40.

Kankaanrinta, I., and T. Masalin. 2001. Groupware—A dynamic internet tool for geography and environmental education. In *Innovative practices in geographical education: Proceedings Helsinki Symposium of the International Geographical Union Commission on Geographical Education*, eds. L. Houtsonen and M. Tammilehto, 115–20. Helsinki: University of Helsinki.

Klein, P. 2005. *CGGE evaluation report*. http://www.aag.org/Education/center (last accessed 20 September 2007).

Kropotkin, P. 1885. What geography ought to be. *The Nineteenth Century* 18:940–56.

Ray, W., and M. Solem. forthcoming. Gauging disciplinary support for internationalization: A survey of geographers in the United States. *Journal of Geography in Higher Education*.

Rediscovering Geography Committee. 1997. *Rediscovering geography: New relevance for science and society*. Washington, DC: National Academy Press.

Reed, M., and B. Mitchell. 2001. Using information technologies for collaborative learning in geography: A case study from Canada. *Journal of Geography in Higher Education* 25(3):321–39.

Reeve, D., S. Hardwick, K. Kemp, and T. Ploszajska. 2000. Delivering geography courses internationally. *Journal of Geography in Higher Education* 24(2):228–37.

Samovar, L., and R. Porter. 2003. *Communication between cultures*. Belmont, CA: Wadsworth Publishing Co.

Solem, M., S. Bell, E. Fournier, C. Gillespie, M. Lewitsky, and H. Lockton. 2003. Using the internet to support international collaborations for global geography education. *Journal of Geography in Higher Education* 27(3):239–53.

INDEX